Feed your children well . . .

Concerned about pesticides in your kids' food? Wonder what you can do to protect your family?

You need good, solid information to understand the problem . . . and a practical plan to solve it. This book offers both.

The time is *now.* Find out about . . .

- Why pesticides are a threat to our kids and our environment
- How to get safe food for your family—wherever you live
- Why many farmers are stopping or cutting their pesticide use—and how you can support them
- How innovative grocers and food companies are offering safer choices to consumers
- What you can do in your own home to protect your family from pesticides in food
- How to write letters, hold meetings, and organize with others for a safer food supply and safer planet

The Way We Grow

Good-Sense Solutions for Protecting Our Families from Pesticides in Food

Good-Sense Solutions for Protecting Our Families from Pesticides in Food

Anne Witte Garland with
Mothers & Others for a Livable Planet

Printed on recycled paper.

THE WAY WE GROW

A Berkley Book/published by arrangement with Mothers & Others for a Livable Planet, Inc., a New York not-for-profit corporation

PRINTING HISTORY
Berkley trade paperback edition/July 1993

Interior design by Laurie Mazur.
Interior illustrations by Li Howard.
Photos: page 5, World Wildlife Fund; page 56, Kevin Neel; page 58, © Clemens Kalischer; pages 59 and 62, NYT Pictures.

ISBN: 0–425–14061–X

BERKLEY®
Berkley Books are published by The Berkley Publishing Group, 200 Madison Avenue, New York, New York 10016.
BERKLEY and the "B" design are trademarks belonging to Berkley Publishing Corporation.

PRINTED IN THE UNITED STATES OF AMERICA

10 9 8 7 6 5 4 3 2 1

We dedicate this book to our children—
Kenji, Gardner, Lucy, Wyatt, and Ryan—
and to all children.

Acknowledgments

This book is the result of a wonderfully cooperative effort of all my colleagues at Mothers & Others for a Livable Planet: Kate Roth, Wendy Gordon, Betsy Lydon, Kristin Ebbert, Pajarita Charles, and Cinnie Cole. Lisa Lefferts, science consultant to Mothers & Others, provided invaluable research and support; I look forward to many more collaborations with her.

The following people kindly reviewed and gave their thoughtful comments on all or portions of the manuscript: Catherine E. Adams, Ph.D., R.D., Director of Scientific Affairs, Grocery Manufacturers of America; Edward Groth III, Ph.D., Director, Technical Policy and Public Service, Consumers Union; Harvey Karp, M.D., pediatrician, Pier Avenue Pediatrics and Assistant Professor of Pediatrics, UCLA School of Medicine; Frederick Kirschenmann, Ph.D., Manager, Kirschenmann Family Farms, President, Farm Verified Organic, and President, Organic Foods Production Association of North America; Lawrie Mott, M.S., Senior Staff Scientist, Natural Resources Defense Council; David Pimentel, Ph.D., Professor, Department of Entomology, Cornell University; and Robert Scowcroft, Executive Director, Organic Farming Research Foundation.

Donald C. Farber, of Counsel to Tanner, Propp, and Farber, made this book possible. Hillary Cige, Jacky Sach, and other staff at The Berkley Publishing Group enthusiastically guided it through production on a very tight schedule. Laurie Mazur and Li Howard cheerfully performed impossible feats to create the design, layout, and illustrations on time. John McNeel was our able and agreeable copyeditor. The Jaffe Family Foundation and the Rockefeller Family Fund provided Mothers & Others with the financial support to research and write the book.

Special thanks to the Mother & Others board of directors: Lynne Edgerton, attorney and Vice President, CALSTART; Wendy Gordon, M.S., Mothers & Others Co-founder, Acting Chair, and Program Director; Harvey Karp, M.D.; Diane MacEachern, President and Co-owner, Vanguard Communications; Lawrie Mott, M.S.; Deborah Schimberg, consultant to Providence Public Schools and Founder and Former Executive Director, Southside Community Land Trust; Jane B. Stewart, Environmental attorney, Paul, Weiss, Rifkind, Wharton & Garrison; Meryl Streep, Mothers & Others Co-founder; and Roberta Willis, Founding member, Connecticut Mothers & Others for Pesticide Limits.

Finally, personal thanks to David and Kenji Garland for their constant patience and support.

Contents

Foreword

by T. Berry Brazelton, M.D.

We are endangering the future health of our children. The information in this book about the hazards of pesticides—and particularly their effect on children—is bound to worry parents. And in fact, all of us *should* be concerned about how our misuse of the environment, including our overuse of pesticides in growing our food, will affect our children's future.

We must not let these problems make us feel powerless. Parents and grandparents are likely to feel angry and guilty because we've allowed our environment and food supply to become so contaminated. But we can do our children even more harm that way. Instead, we need to look for short-term and long-term actions that we can take to protect our children. With action, we can feel empowered, and our children will see that we can and will have an effect on our world.

As a pediatrician and a grandparent, I share many of parents' concerns about our children. This book is about three of our greatest concerns: our children's food, our children's health, and our children's future. *The Way We Grow* makes me hopeful, not fearful. It is a positive, practical book that tells us how to turn our concerns into action to protect our children and their future.

Right now, too many pesticides are used to grow the food in our country. This is hurting the environment our children live in and will inherit. It is making farming unsafe for farmers and farm workers. And it is contaminating the food supply. Fortunately, though, there are solutions.

Four years ago, I joined with Mothers & Others to call on parents to work together to solve the problem of pesticides in our children's food. We have made considerable progress since then. Many farmers have reduced their use of chemicals in growing food, and thousands of others have stopped using pesticides altogether. These are very promising signs, but there is still much to do. The government needs to institute far better pesticide controls, and to encourage and help many more farmers to switch away from chemical-intensive farming.

The Way We Grow shows us step-by-step how to change the food production system to protect our children and the environment. It lays out clearly and logically what each of us has to do—from feeding children healthful diets with plenty of fresh fruit and vegetables, to working to make safer food available for everyone, to pressing for important political reforms.

The changes that Mothers & Others calls for in this book are fundamental changes that will mean creating a new food system in this country. These changes will threaten the profits of the chemical industry, and we can expect opposition. Achieving the reforms will require a massive public effort to overcome that opposition. We must all work *together*—parents and grandparents and others who are concerned about children, citizens and consumers and environmentalists, farmers and food retailers. *The Way We Grow* is a model for cooperative activism for safer food and a saner food policy. I urge you to read it and start today to protect your family, and to work with Mothers & Others for a new food system, for our children's future.

Introduction

Steamed broccoli with dinner . . . the banana or peach packed in your child's lunch box . . . a bowl of breakfast cereal or slice of whole-wheat toast . . . Fresh or frozen, packaged or processed, the food you feed your family comes *from the earth*. If you're like most Americans, as you push your cart down the supermarket aisle every week or so, you're probably wondering just how safe your food really is.

Were chemicals used to grow that broccoli? Was your child's peach or banana sprayed with fungicides? What kinds of pesticides were used to grow your food? Are these chemicals still on the food when you buy it? Are they safe? Can they be washed off or peeled away?

Are they in the food itself? How many pesticides will your family be eating with tonight's dinner? What does this mean for your health—*or your child's?*

As a consumer and caring parent, you clearly have a stake in how your family's food is grown. Mothers & Others for a Livable Planet has written this book to show why you should care about the way we grow our food in America, how our food production system needs to be changed, and—most importantly—what you can do, along with Mothers & Others, to make it better for your family and your community.

It *is* time for a change. The industrial, chemical-intensive agriculture that developed in this country after World War II made it possible to produce abundant, cheap food, but the over-reliance on farm chemicals has exacted a high price. The health and environmental costs of conventional agriculture are hidden costs that don't show up at the supermarket cash register—but they're real costs, and *serious* ones. The overuse of chemical pesticides has compromised the safety of our food and put our children's health at risk. Synthetic pesticides and fertilizers have polluted our water and soil, damaged agricultural and natural ecosystems, and caused serious health problems for farmers and farm workers. Chemical-intensive agriculture has developed in a way that has hurt rural communities and family-run farms; it has contributed to forcing many farmers off the land while concentrating considerable wealth and power over our food supply in the hands of a few of the nation's largest corporations.

For the sake of our kids, for the sake of our environment, for the sake of farmers, we need to reverse the trend of using more farm chemicals—and instead adopt a safer, sounder system of agriculture.

Obviously, it can't happen overnight, but it *can* happen. The problems are serious, but the solutions—a surprising variety of them—abound. Beginning today, you can make important changes in the way you shop for the food that you feed your family—simple, practical changes that we outline for you in this book. With these changes, you'll be protecting your children right now from potentially dangerous pesticide residues in their food, while voting with your dollars for a saner agriculture policy. You will become more connected to our food production system—better informed about where your food comes from, and more empowered to say how it should be produced. And you'll be working to ensure the integrity of the food production system for generations to come, by helping to provide farmers with the support they desperately need to grow food safely and in a way that safeguards the environment.

The best farmers have traditionally been guardians of the land—observing, nurturing, working the earth to keep it healthy, fertile, and productive. It's only been in the past several decades that much of our food production system has shifted to non-sustainable farming methods—a "bigger is better" approach that emphasizes maximizing yields and relies increasingly on costly chemical fertilizers and pesticides for quick fixes to soil and pest problems. Unfortunately, federal and state agriculture policy and financial lending practices support shortsighted farming techniques, and create difficult obstacles for farmers who want to adopt practices that would be safer for the environment, the food supply, and farmers themselves. In spite of the obstacles, though, more and more farmers are choosing to get off the pesticide treadmill and are doggedly proving that safe, "sustainable" farming is possible—and profitable. It's time to overhaul our governmental and financial policies to give all farmers the support they need to take care of the land.

Mothers & Others for a Livable Planet has been working on the issue of pesticides and food safety since early 1989. That year, the Natural Resources Defense Council (NRDC) published an influential report, "Intolerable Risk: Pesticides in Our Children's Food," which documented how children are being *legally* exposed to pesticides in their diets at levels above what the government *itself* considers safe. Out of concern about the information in NRDC's report, Meryl Streep, Roberta Willis, and some of their neighbors in a small Connecticut town organized the first local Mothers & Others group to start educating themselves about pesticides in food and to find safer alternatives. Based on that model of parent activism (described on page 59), the national Mothers & Others was launched—initially as a project of NRDC, and now as a fully independent non-profit organization—to inform people about the risks of pesticides in food, and how to protect their children and press for needed reforms.

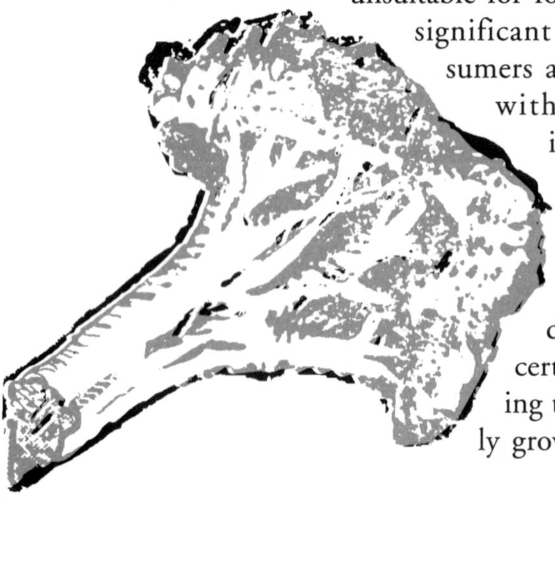

Parents were eager to join our efforts. Mothers & Others' first book, *For Our Kids' Sake: How to Protect Your Child Against Pesticides in Food,* helped raise the issue nationally. Mothers & Others helped focus public scrutiny on one particularly hazardous chemical, Alar, which has since been banned by the Environmental Protection Agency (EPA) as unsuitable for food uses. The issue sparked significant new interest among consumers and farmers in food grown with reduced pesticide use, including food grown organically. In 1990, Congress enacted the first-ever national organic standard, to provide a clear definition of organic, and a certification process for ensuring that food labeled "organically grown" truly is. That national standard and label will help set the stage for major expansion of organic farming and the organic industry, and will make it easier for consumers to choose food grown without synthetic pesticides. It should also help foster more interest, among consumers and farmers alike, in the overriding benefits of sustainable agriculture.

Buying certified organically grown food is just one of many ways for us to protect our families from pesticides in food and to support sound agricultural practices. With enough consumers exercising our considerable buying power, we will send a clear message that we want food that is grown safely and sustainably—for the environment and for our kids.

How to use this book

Mothers & Others receives frequent calls and letters from parents who want to know how to protect their kids from pesticides, where to find organic food, and how to tell where their fruits and vegetables come from and what's on them. *The Way We Grow: Good-Sense Solutions for Protecting Our Families from Pesticides in Food* answers those questions and more.

The Way We Grow draws on Mothers & Others' experience of the past four years, working day-to-day on the issue of pesticides and food safety with parents and parent organizations, consumer and public health groups, farmers and farm worker groups, and retailers and other industry representatives. *The Way We Grow* details simple and effective steps that you can undertake on your own, or together with other concerned parents in your community. We know very well how busy parents are—so we have organized *The Way We Grow* to make it easy for you to read, and easy to start today to protect your family and put your buying power to work for a safe and sound food policy.

Nothing inspires the way success inspires, so you'll find plenty of success stories throughout this book—stories of farmers who are rejecting conventional, chemical-intensive practices, and are growing their crops without synthetic chemicals or with dramatically reduced pesticide use; stories of supermarkets that are giving consumers useful information and real choices in the food they buy; stories of giant food-processing companies that are conscientiously choosing to use food grown with fewer pesticides; stories of successful local organizing efforts for safer food. Every time a farmer rejects chemical-intensive farming, it's an act of redeeming the land for future generations. All of these innovators prove that we can have a better food production system, and they point the way—the ways—for us.

Chapter 1, "From the Farm to Your Table," details why we need to change our current food-production system—including health problems for farmers and farm workers, the environmental destruction wrought by chemical-intensive farming, and the health threats to our children because of pesticides in their food. In addition, this chapter details the ways in which government pesticide regulations and farm policies have failed to protect our children and the environment. Chapter 2, "A New Food System," tells how positive change is possible—by relating some very promising developments in food production, such as the increased market in organically grown foods, widespread new interest in reducing pesticide use, and innovations in processing and retailing.

If you're most interested in practical advice, you may choose to read these two chapters later, and instead first skip ahead to Chapter 3, "What You Can Do." This chapter begins with ways for you to protect your family right now from potentially harmful pesticide residues in food. It also tells you about sources of food grown with fewer or no pesticides, how to increase your access to such food, and ways that food can be more affordable for all families.

A new food system will have to be supported by new government policies. Chapter 4, "Tools for Organizing and Learning More," is a primer in organizing and advocating critical reforms. That chapter includes Mothers & Others'

"Agenda for a new food system"—a point-by-point description of the important changes that are needed to turn our food-production system around. We make it easy for you to take the first step to press for those reforms; at the back of *The Way We Grow* we have included tear-out postcards for you to sign and mail to key congressional decision-makers urging them to support a new food system.

At the end of Chapter 4, you'll also find several pages of information and letters for you to photocopy and use—including letters to President Clinton, EPA Administrator Carol Browner, and Agriculture Secretary Mike Espy, urging them to provide crucial leadership to move our country to a safer and more sustainable food system. In addition, we have included a tear-out check-list of what you can do now to protect your family from pesticides in food and to make your buying power count for a better food system—we suggest that you hang this in a prominent spot in your kitchen or pantry. Or, share the information with other parents in your community by putting it up at your child's day care center or nursery school.

One of the most important things you can do right now is to join Mothers & Others for a Livable Planet (you can use the tear-out form at the back of the book). Mothers & Others is the first and only national organization working exclusively on environmental issues affecting children. Through our extensive public education, organizing, and advocacy work, we speak *to* and *for* parents, on important issues such as pesticides in our children's food. We want to make it easier for parents to make a difference—for our kids' sake. Please join us.

1

From the Farm to Your Table

What's Wrong with the Way Our Food Is Grown

Every year, more than *800 million pounds* of pesticides are used to grow food in this country—dusted on crops, poured into the soil, coated on seeds, and sprayed onto fields from tanks or foggers, planes or helicopters. And they work, at least up to a point: Farm pesticides decrease labor costs, and help increase crop yields by reducing losses to some (though not all) insects, disease, or weeds. They increase the storage life of some food. And they can improve the *appearance* of food so that it can meet marketing standards and command a higher price on the market.

But all of this pesticide use comes with its own enormous price tag. Farmers spend about $6 billion a year directly, and other, hidden costs—including environmental damage and health problems—add up to an additional *$8 billion.* From the farm, to the environment, to our tables, a look at the consequences of our overuse of pesticides repeatedly begs the question, "Is this any way to grow our food?"

Health problems for farmers and farm workers

The people who are most exposed to agricultural pesticides are the farmers and farm workers who work directly with pesticides: Mixing pesticide solutions, applying pesticides on the fields, cleaning out the spraying equipment, working in fields treated with pesticides, and picking and handling the treated

The pesticide price tag

It is impossible to put an exact dollar value on a human life lost or wildlife destroyed because of pesticide poisoning. But for the past several years David Pimentel, a Cornell University entomologist, has taken on the monumental task of calculating the indirect, or hidden costs of pesticide use—costs that don't get taken into account in the usual cost/benefit analyses conducted by industry or government regulators. In 1992, Pimentel estimated that the environmental and social costs from pesticides in the U.S. total more than *$8 billion* a year:

Public health impacts	$787,000,000
Domestic animal deaths and contamination	30,000,000
Loss of natural enemies	520,000,000
Cost of pesticide resistance	1,400,000,000
Honeybee and pollination losses	320,000,000
Crop losses	942,000,000
Fishery losses	24,000,000
Bird losses	2,100,000,000
Groundwater contamination	1,800,000,000
Government regulations to prevent damage	200,000,000
Total	**$8,123,000,000**

crops at harvest time. What's more, the people who live on or near farms drink well water that is often polluted with pesticides and other agricultural chemicals. They're exposed when nearby crops are sprayed and the wind blows pesticides to their homes, to the fields where they're working, or to areas where their children are playing. Workers bring pesticides home from the fields on their clothes and shoes. Children may swim or play in streams or other water that has been polluted with agricultural chemicals. Farmers who have switched from conventional farming to organic farming often cite concerns about their own children's exposure to pesticides as a reason for the switch.

Marion Moses, a physician and advocate for farm workers who has done much of the work documenting health problems of pesticides, points out that because of a lack of child care, the young children of farm workers often go with their parents into fields, where they're directly exposed to pesticides. When pregnant women work in pesticide-sprayed fields, their fetuses are exposed. Birth defects have been documented among babies born to women farm workers, and according to Moses, the reason even more birth defects haven't been found may be that pesticide exposure causes miscarriages and still births. The health problems are often compounded by a lack of adequate, clear information about pesticide hazards, and a lack of information and education about symptoms of pesticide poisoning—for farmers and farm workers themselves, as well as for the physicians who treat them.

Pesticides are *poisons*—to pests *and* humans. Their acute health effects—effects that appear within hours or days of exposure—can range from skin rashes and eye irritation to respiratory problems, systemic poisoning, and death. The chronic, or long-term, effects of pesticide exposure may include cancer, nervous system damage, birth defects, and reproductive and fertility problems. A National Cancer Institute study found that children living in homes where household and garden pesticides are used are seven times more likely to develop childhood leukemia than other children. Studies have shown that farmers have higher-than-normal rates of several kinds of cancers, including leukemia, multiple myeloma, non-Hodgkin's lymphoma, and cancers of the brain, prostate, stomach, skin and lips.

Sylvia Ehrhardt, who grows fruit and vegetables organically on a 76-acre farm in Maryland, has a graphic way of driving home how hazardous pesticides are. She hands visitors to her farm a copy of pages from a trade catalog for farmers advertising protective clothing to wear while working with pesticides—including nylon and vinyl spacesuit-like coveralls, protective hoods, masks, and gloves. In one picture, under the headline, "Why you need protection," arrows label different parts of the body with the greatest susceptibility to chemical absorption—in this case, the chemical parathion: the ball of the foot (13.5-percent absorption), abdomen (18.4 percent), scalp (32.1 percent), forehead (36.3 percent), ear canal (46.5 percent), and scrotum (100 percent). Another photo shows a man wearing full protective gear, standing in the middle of a field.

Is this any way to grow our food?

Pesticides and the environment

When pesticides are used in farming, they end up in our soil, air, and water supplies. Rachel Carson was one of the first to publicly sound the alarm about pesticide use and its effects on the environment, in her 1962 book, *Silent Spring.* In it she warned that after what was then less than two decades of use of synthetic pesticides in agriculture, pesticides such as DDT were already present virtually everywhere in the environment. DDT is so persistent that even though it was banned in the U.S. in 1972, DDT and its breakdown products are still found in soil, in lakes, and in ten percent of the food tested annually by the U.S. Food and Drug Administration. What's more, they are present in the tissues of animals (including in areas of the world where DDT was

Profile of a scientist: Theo Colborn

Theo Colborn is not your stereotypical scientist, pursuing her research in an ivory tower. Instead she's an activist and pioneer who is making important connections between effects of pesticides on wildlife and the implications for people.

Colborn's work studying wildlife in the Great Lakes area has uncovered alarming effects of pesticide and chemical poisoning of the environment—including reproductive abnormalities, birth defects, and damage to hormonal, immune, and nervous systems. As a result of what she and her colleagues have found, Colborn has become a vocal critic of the way the government regulates pesticides. As she has put it, "Ask blundering questions. Provide weak answers. Generate bad policies." The blundering questions include an overriding focus in current testing and regulation on only narrow effects of pesticide exposure, particularly cancer, when other, equally serious effects may be much more pervasive and may result from lower exposures.

Colborn doesn't claim to have all the answers. She admits that there are uncertainties in drawing conclusions about the human health effects of pesticide exposure from its effects on wildlife. But she isn't afraid of looking at all the danger signs and asking the difficult questions, and insists that government and industry must ask them, too. ❧

never sprayed), as well as in the blood and tissues of humans—including many Americans.

Today, the degradation of natural and agricultural ecosystems caused by the overuse of pesticides is well-documented, by experts like Cornell entomologist David Pimentel, who has done groundbreaking work in calculating the environmental and social costs of pesticides (see "The pesticide price tag," page 2). Among other problems, the toll on the environment and on wildlife includes:

➤ **Soil degradation.** Pesticides and other agricultural chemicals make it possible to practice "monoculture"—planting the same crop in the same fields year after year, instead of rotating crops to rebuild soil nutrients, control erosion, and thwart pests that thrive on one kind of crop but are deterred by others. Monoculture has weakened soil and led to increased fertilizer use and massive soil erosion; it's been estimated that at least one-third of the topsoil in this country has been lost because of the switch to chemical-intensive agriculture, and an average of about seven tons per acre continue to erode every year.

➤ **Damage to crops and to beneficial insects.** Not all pesticide use benefits crops; pesticides destroy some crops as well, either through normal use or accidentally through excessively high doses or spraying of adjacent fields. Pesticides are also indiscriminate in their destruction—they can and do wipe out not just the harmful pests, but beneficial insects and microorganisms that otherwise would help keep pests under control. So pesticides can lead to a "rebound" effect—wiping out natural ene-

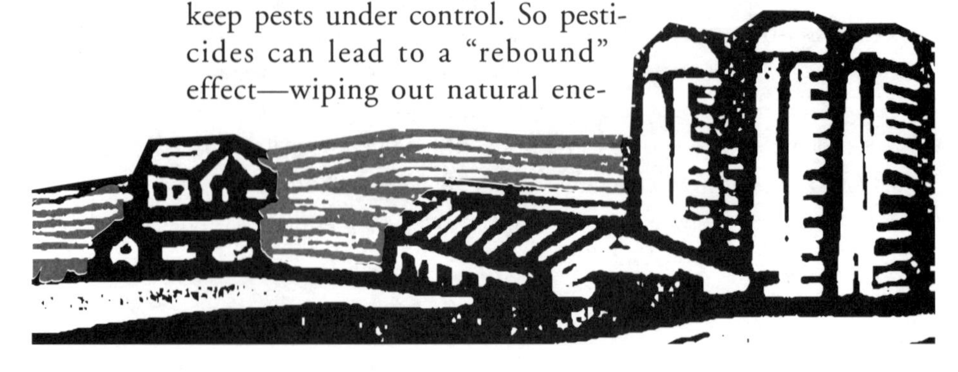

mies and allowing pest populations to return with a vengeance. In addition, pesticides can aggravate problems with "secondary" pests. These are pests that initially aren't present in high enough numbers to do much damage, but when their natural enemies are destroyed, secondary pest populations grow rapidly.

➤ **Pesticide resistance.** In our battle against pests, the pests are winning. Because of extensive pesticide use, more than 900 insect, plant pathogen, and weed species have developed a resistance to one or more pesticides—which leads to increased pesticide use or the need to develop new, often more toxic, pesticides. And the vicious cycle continues, until pests develop a resistance to the *new* pesticides. Since 1945, pesticide use on farms has increased more than 30-fold, but at the same time the portion of total harvest lost to pests has still increased by 20 percent.

➤ **Harm to bees and wildlife.** Pesticides have decimated huge numbers of honeybee and wild bee populations—populations that are critical for the pollination of many fruit crops. In addition, pesticides and other agricultural chemicals have wreaked havoc with certain wildlife populations. Pesticide contamination of streams and lakes, fields, and other fish and wildlife habitats kills millions of fish, birds, and other wildlife every year either directly or by destroying or poisoning their food supply. Wildlife populations have been affected in other ways as well—including decreased fertility, birth and behavioral abnormalities, and compromised immune systems.

➤ **Contamination of water supplies.** The manufacture and use of pesticides has seriously polluted our water supplies. Because of the proximity to agricultural areas, as much as half of the groundwater and half of the well water in the U.S. is at risk from being contaminated by farm pesticides. About half of the people in the country rely on well water for their drinking water. The EPA has found 98 different pesticides in the groundwater in 40 states—contaminating the drinking water of more than 10 million people. Once pesticides get into groundwater, they're virtually impossible to remove, and they can persist there for decades.

Is degrading our environment any way to grow our food?

Pesticides in our children's food

Eventually, pesticides sprayed on food crops make their way onto our food. Although the amounts are usually quite small by the time they reach consumers, *any* amount of some of these poisons may be cause for concern—particularly the pesticides that can cause cancer or affect the nervous system, and particularly for children. Scientists have found it very hard to quantify the risks from exposure to pesticides in food, but to the extent that these risks are unnecessary and involuntary, *there is no reason to take them.*

Of about 300 pesticides approved by the federal government for use on food crops, 73 are "probable" or "possible" carcinogens (cancer-causing substances), including some of the most frequently used pesticides. (The EPA calls a substance a "probable" carcinogen when there is limited evidence from human epidemiological studies or strong evidence from animal studies that the substance can cause cancer. If there is limited evidence in animals and no human data, EPA calls the substance a "possible" carcinogen.) The four most commonly used pesticides in the U.S. (atrazine, alachlor, metalochlor, and 1,3-dichloropropene) are all classified as either probable or possible carcinogens, and of the 20 most frequently used pesticides, seven are potential carcinogens.

Although much of the study of risks associated with pesticides has focused on cancer, it isn't the only health problem associated with the kinds of pesticides used in growing our food supply. Classes of pesticides known as organophosphates and carbamates (which were substituted for DDT and similar pesticides when they were banned) are "neurotoxins," meaning they can cause nervous system damage. Some pesticides can cause genetic or reproductive damage, some disrupt the body's hormonal system, and some can harm the body's immune system.

Children may be more vulnerable than adults to the effects of pesticide exposure—including to pesticide residues in their food—for a number of reasons:

➤ Children eat relatively more food, particularly fruit and vegetables, than adults do—so they receive greater exposure to pesticides that are on that food. Non-nursing infants consume 15 times more pears, and 16 times more apple juice, than the average adult, for instance (per pound of body weight). That means they're exposed to any pesticide that may be on those foods at a rate 15 and 16 times higher than average.

The circle of poison

It can take *years* for the federal government to ban a dangerous pesticide, even when the government recognizes that the pesticide could pose a significant health risk. When a pesticide finally *is* banned, what happens next? One scenario goes like this: the chemical company just keeps on making it (endangering chemical workers and polluting our environment in the process), the pesticide gets exported to a country with laxer laws, it is used on food crops there, the food is imported into the U.S. (in the winter, almost half of the produce consumed in this country is imported), it gets past the FDA's monitoring (only about one percent of imported food is tested), and the banned pesticide ends up on our tables anyway. This is the "circle of poison."

It doesn't just happen with banned pesticides, but also with "unregistered" pesticides that were never approved for use on food crops in the U.S. to begin with. The practice turns other countries into dumping grounds for pesticides considered too dangerous for use here, and puts agricultural workers in those countries at considerable risk because they may not be adequately equipped with training or protective garments.

Serious reform of pesticide regulation has to include ending the circle of poison, by stopping the export of any unregistered or banned pesticides. Predictably, though, U.S. chemical manufacturers have lobbied hard to protect their profits; Congress needs to hear from the public that the circle of poison must be stopped now. ❧

➤ A young child's "detoxification" system isn't fully developed. A newborn's kidneys, for instance, are immature compared to an adult's, which makes the infant more vulnerable to chemicals that depend on the kidney to be eliminated from the body. Certain enzymes that help to detoxify many chemicals aren't fully functional or aren't present at all in young children, or are present at lower levels than in adults. A child's metabolism is also different from adults. While this makes children less vulnerable than adults to some toxins, they are *more* vulnerable to others.

➤ A young child's brain, nervous system, immune system, and other systems and organs are still developing and are more vulnerable than an adult's.

➤ When children are exposed to toxins, there is more time for resulting damage to develop than when adults are exposed. Children have their whole lives ahead of them. If a series of events has to occur before toxic effects of a chemical are manifested (for instance, the process by which a normal cell is transformed into a cancerous cell), then it's more likely that those events *will* some day unfold if the exposure takes place at age one than, say, at age 65.

➤ Children appear to be more susceptible to some carcinogens than adults are. In animal testing of carcinogens, animals exposed to carcinogens around the time of birth often (but not always) develop more tumors, and within a shorter time, than animals exposed as adults. And children are definitely more sensitive to many pesticides that disrupt the nervous system.

Clearly, if we want to protect children, then it is simply being prudent to reduce their exposure to pesticides as much as possible. *Is risking our children's health any way to grow our food?*

What the government is (isn't) doing

Unfortunately, in regulating pesticides in food, the federal government has in large part ignored children's eating patterns and unique susceptibility to toxins, and has instead treated them in effect as mini-adults. (And even adults may not be adequately protected.) To make matters worse, current federal agricultural policy actually promotes the overuse of pesticides in farming, by making it difficult for farmers to adopt farming practices that would require fewer or no chemicals, and by failing to make widely available the information and support farmers need to switch to less chemical-intensive methods.

Regulatory problems. Two agencies are largely responsible for federal pesticide regulation. The EPA regulates pesticide use—approving pesticides for use on food crops (with the mandate of preventing "unreasonable adverse effects" on the environment and public health), and setting legal limits for pesticide residues in food. The Food and Drug Administration (FDA) is responsible for monitoring food for pesticide residues—except for meat and poultry, which are regulated by the U.S. Department of Agriculture (USDA)—and making sure that food contaminated with illegal levels of pesticides doesn't reach consumers. Because of weaknesses in

Beauty that's only skin deep

If you've always been careful to buy only perfect-looking produce, here's a reason to think twice next time you shop: It may have taken a lot of pesticides to achieve that perfection.

To get top dollar for their products, farmers often have to meet strict cosmetic quality standards, including USDA grade standards and separate standards set by packing houses or food processors. Meeting those standards may require extensive pesticide use, however; David Pimentel of Cornell University estimates that from 10 to 20 percent of insecticides and fungicides are applied simply to comply with strict cosmetic standards.

Since retailers believe that consumers *want* perfect-looking fruit and vegetables, let your grocer know that you would rather see a few blemishes, if that means that fewer pesticides are used to grow your food. ❧

the law itself and the way the agencies carry it out, current pesticide regulation is riddled with flaws:

➤ *Federal pesticide policy works to favor the most toxic pesticides already in use over newer, potentially more benign pesticides and non-chemical pest-control methods.* When the EPA approves, or "registers," a pesticide for use on food crops, it does a cost/benefit analysis comparing the costs of using that pesticide with its economic benefits, such as projected increases in crop yields. But these assessments haven't been well-balanced: They have favored the economic benefits of pesticides already in use, without taking into account non-chemical or lower-chemical pest-control alternatives. In addition, they rarely account for all of the costs, including the pesticides' full impact on the environment. Even when the EPA has some evidence that a pesticide is hazardous, it can take years before it is removed from the market, while the agency conducts a "special review" of the pesticide.

➤ *Most pesticides haven't even been fully tested for their health effects.* Instead, they were registered before current laws were in place to require such testing. The EPA is now in the process of re-registering older pesticides, and is requiring more testing data as part of that process. But the process is proceeding very slowly (and the pesticides stay on the market in the interim)—of more than 400 pesticides, only 29 had been re-registered as of December 1992. Meanwhile, a number of scientists are concerned that even the level of testing of pesticides that is now required is inadequate for certain very serious health effects, including their impact on the immune, nervous, reproductive, and hormonal systems—problems of particular concern for children.

➤ *In setting the maximum legal limits (or "tolerances") for pesticide residues on food, the EPA doesn't consider all exposures to pesticides.* Tolerances are set for individual pesticides, on individual crops. In practice, several pesticides might be present on any one crop, but the EPA doesn't consider the combined effect of exposure to those pesticides, or of pesticides that might be on other food. People can be exposed to pesticides in many other ways as well, such as pesticides used in homes or schools; chemicals used on lawns, gardens, or parks; routine spraying in communities for mosquitoes; or pesticide drift from aerial spraying—none of which is taken into account in tolerance-setting for food-use pesticides. The EPA also doesn't consider the "inert" ingredients in pesticides (ingredients other than the "active" ingredients that actually kill the pests), even though some inert ingredients may be toxic themselves, or may be added to a pesticide product to boost its toxicity.

➤ *The EPA has based tolerances on outdated and possibly inaccurate information about Americans' food consumption habits, which may underestimate exposure to pesticides—particularly children's.* In the process of re-registering pesticides, the agency is updating the tolerances of older pesticides, but even the new tolerances are often based on food consumption data from the late 1970s.

➤ *The FDA's monitoring and enforcement programs are very limited, and don't always keep contaminated food from reaching consumers.* The FDA samples only a tiny portion of food—less than 1 percent—to test for illegal pesticide residues. What's more, the testing methods that the agency uses most commonly cannot even detect about *half* of the pesticides approved for use on food. Even when the agency does find illegal pesticide residues on food, it is not always kept from the market, in part because the penalties for distributing contaminated food are not high enough to always be an effective deterrent. And in some cases, by the time the FDA has tested a sample and found illegal residues, the food shipment has already reached the market and been sold to consumers.

Obstacles to cutting pesticide use. Considering the problems associated with overusing pesticides and regulating them effectively, it would clearly make sense for the federal government to embrace and wholeheartedly promote a safer agricul-

Future food?

Juicy, flavorful tomatoes year-round. Corn and peas with longer shelf life and enhanced natural sweetness. Insect- and disease-resistant produce that can be grown with fewer pesticides. These and other unnatural wonders are some of the promises of biotechnology—the process of using living things or components of living things to produce food, food additives, drugs, and other products. Genetic engineering and other forms of biotechnology are extremely powerful tools that *could* be used to complement a new, safe and sustainable agriculture—if, for instance, they were used to develop safer substitutes for hazardous pesticides or other agricultural chemicals. On the other hand, we should be cautious. Biotechnology could be used to concentrate even more power over our food supply in the hands of a few giant companies, giving farmers and consumers less say over what food is grown and how. And in fact, the companies that are bringing us biotechnology are many of the same ones that got us onto the farm-chemical treadmill. So it's no wonder that some companies are using biotechnology to develop plants that are tolerant to the pesticides they manufacture.

With biotechnology, we need a prudent policy of effective regulation to protect consumers and the environment. The government has so far taken a "hands-off" approach: In a 1992 policy statement, the FDA took the position that most genetically engineered foods should be considered safe and don't require the kind of safety testing used for food additives, and that bioengineered food meeting certain criteria should generally not have to be labeled as such. The Clinton Administration could introduce a new approach to regulating biotechnology, however. Vice President Al Gore argued back in 1991 for a rational, unified federal regulatory system for biotechnology, "to provide safeguards against illegal and unethical uses . . . [and] to encourage uses of biotechnology that enhance the economic viability of local communities and interests." ❧

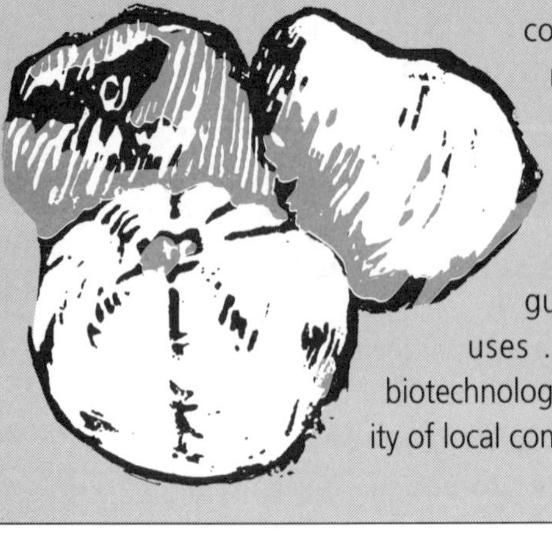

ture policy. But instead of actively helping farmers who want to make the transition from chemical-intensive to lower-chemical or organic farming, the government actually can make it more difficult. The lack of commitment in federal farm policy to alternative agriculture is evident in the budget for the USDA. The budget for all USDA research and technical assistance programs supporting sustainable, lower-chemical farming represents less than 5 percent of the USDA's total budget for research and technical assistance.

Switching from conventional farming to methods that rely less on pesticides and other chemicals has paid off for many farmers, but no one claims it is *easy,* especially in the short term. Farmers who use conventional methods have geared some of the most basic decisions about their farming operations to the routine use of chemicals—decisions as basic as what crops to grow, from which seeds and on which fields, when to plant and spray and harvest, what equipment to buy and how to use it, and how to market the food.

Starting to farm more sustainably doesn't mean simply stopping chemical use; it requires fundamental changes in planning and farming practices. Many farmers are barely making ends meet, and can't afford to take the risk that goes with such basic changes without support, including reliable information, technical assistance, and economic incentives. Vast federally funded programs are in place that *could* conduct valuable research and provide farmers with needed information and technical support, but existing federal and state research and the USDA extension service network focus more on conventional, chemical-intensive farming methods than on alternatives.

Some innovative cooperative extension service programs are providing real services for farmers who want to reduce pesticide use—the University of Iowa in Ames is cooperating with the Practical Farmers of Iowa in on-farm research into farming with fewer chemicals, for instance—but such programs are in a minority. For many farmers, extension services are a main source of technical information—other than pesticide and fertilizer sales people.

Far from getting the *financial* support they need to reduce pesticide use, farmers switching from conventional agriculture are sometimes penalized under federal programs. Federal farm support programs reward farmers for maximizing their yields (in what Iowa farmer Sharon Thompson calls "produce, produce policies")—even at the expense of the environment and the long-term viability of their farms. To reduce pesticide and other chemical use, for instance, farmers have to move away from monoculture and instead rotate crops on fields to help conserve soil, restore nutrients, and break pest cycles. But payments to farmers under federal farm commodity programs are tied to the acreage that is planted with a particular crop; when they rotate crops, farmers enrolled in commodity support programs may end up forfeiting some of their federal payments. And the major commodity markets further encourage monoculture.

USDA grading standards may discourage farmers from cutting chemical use, too, since they require food to meet certain standards of appearance—standards that have nothing whatsoever to do with taste or nutrition—and farmers often apply pesticides to crops just to meet those standards (see "Beauty that's only skin deep," page 11).

The effect of the federal government's pro-chemical mentality is evident in bank lending practices as well. Banks often won't give loans to farmers who don't participate in federal farm support programs. And farmers who are trying to eliminate or reduce their use of pesticides often cannot get credit; the banks say they simply can't take chances on someone who isn't using pesticides.

Considering the obstacles, it's a wonder that any farmers at all are making the switch from conventional farming. Fortunately, large numbers of farmers are, or are ready to.

2

A New Food System

Good News for Farmers, the Environment, and Your Family

Dick and Sharon Thompson's farm in Boone, Iowa, is approximately 300 acres—about average for farms in Iowa. The size may be the only thing that's typical about it, though. At a time when most Iowa farmers grow just two crops—corn and soybeans—the Thompsons raise corn, soybeans, oats, hay, and livestock. On a typical farm, chemical fertilizers would be used to put nutrients into the soil, herbicides would be applied regularly to kill weeds, and antibiotics might be routinely added to livestock feed.

Dick and Sharon Thompson

That was how the Thompsons farmed as well, until 1968, when they quit using farm chemicals.

"We were 'high-input' farmers from 1957 to 1967, and purchased everything the salesmen had to sell," says Sharon. "Dick was building his kingdom, where enough was never enough. We changed to balanced farming in 1968 and shut the door on all those salesmen." For the past 25 years, the Thompsons have experimented, studied, made mistakes, adapted, had successes, and kept on experimenting. Today they farm almost exclusively without chemical pesticides, using herbicides only in emergencies. The yields they're able to produce compare favorably with those of the most successful conventional farms. "Our experiences tell us that there are no quick answers or no one special recipe to solve agriculture's problems," says Sharon. "We find ourselves asking more questions each day, and hope we are asking the right questions."

To find answers, the Thompsons conduct extensive on-farm research on methods for reducing chemical use—including research that can benefit other, more typical farms that aren't diversified, for instance. They regularly share their experiences ("5 percent inspiration and 95 percent perspiration," they say) with other farmers and researchers; since 1986, more than 27,000 people from across the country and around the world have visited their farm at annual farm tours. The bottom line, as Sharon has put it, is "profitable and environmentally sound farming—pure and simple. It's got to sustain the land, the soil, the people, the communities, and the pocketbook. It all has to fit together."

It *is* fitting together, for more and more farmers. Now it's up to consumers to show that we care about the changes that are being made and the ones that are still possible.

Toward sustainable agriculture

The kind of farming that the Thompsons and tens of thousands of other farmers are practicing today is, very basically, "sustainable" farming—farming that *sustains* the health of farmers themselves, the economic viability of their farms and farming communities, the life of the soil, the quality of the environment, and the abundance and safety of the food supply. Within this broad notion of sustainable farming are many approaches, ranging from Integrated Pest Management, which involves the minimal and judicious use of chemicals, to organic farming systems that rely almost exclusively on natural materials.

Sustainable farming practices include a variety of specific methods to conserve soil and water and to resist pests naturally, while avoiding or reducing the use of chemicals. The methods or combinations of methods used vary among farms and among regions, according to factors such as climate, soil, crop types, and farm size. Sustainable methods often include using crop rotation and soil-conserving tillage practices, intensively monitoring for pests and disease, timing the planting and harvesting of crops to avoid pest damage, using biological controls that assault pests with natural predators, or planting cover crops or crop varieties that resist insects and weeds.

The notion of sustainable farming isn't pie-in-the-sky; it is practical *and* necessary. "We have said for years that balanced or sustainable farming practice was a viable alternative in agriculture," says Sharon Thompson. "Today we have to say that it is the only common-sense solution to the problems facing agriculture and mankind."

Numerous studies have shown that overall pesticide use in the U.S. could be cut by 35 to 50 percent—without decreasing crop yields, and with just a one percent increase in the retail price of food. Experiences here and in other countries show that reductions like that are not just *possible*—they are already being accomplished.

How was that food grown?

Throughout *The Way We Grow* we refer to "sustainable" agriculture. Unfortunately, you can't walk into your supermarket and find food labeled "sustainably grown." However, you may find—or you can ask for—food grown using several sustainable farming methods. Right now, the *only* sustainable farming approach with a national definition and certification program is **organic** farming. (Organic food is produced using sustainable production methods that rely primarily on natural materials.) We need clear, enforceable national definitions, standards, and third-party certification programs for other terms as well. Meanwhile, here are widely accepted definitions for these terms (they are described in more detail throughout the book):

➤ **Transitional organic** means that the farmer is using organic methods, but has not yet been certified as organic.

➤ We consider **locally grown** food to be grown sustainably, since it is less likely than other food to have been treated with post-harvest pesticides, and since less energy is consumed transporting this food long distances.

➤ Food labeled **"no detected residues"** has been tested for any pesticides used by the grower, plus a variety of other pesticides, and found to contain no detectable pesticide residues above 0.05 parts per million.

➤ **Integrated Pest Management (IPM)** is a pest population management system that prevents pests from reaching damaging levels by using techniques such as natural enemies, pest-resistant plants, cultural managment, and minimal and judicious use of pesticides. ❧

The promise of Integrated Pest Management

In the 1970s and early 1980s, the Indonesian government learned the hard way about some of the problems of overusing pesticides. The country had been importing more and more of its supply of rice, and in the pursuit of self-sufficiency, the government initiated a massive chemical-intensive farming program that included, among other measures, the use of high-yield rice varieties, the elimination of crop rotations, and subsidies for pesticides and fertilizers. But the initiative backfired, when the extensive, routine use of insecticides wiped out the natural predators of a particularly destructive pest, the brown rice planthopper—which flourished and by 1985 threatened 70 percent of the country's rice crop. A pilot Integrated Pest Management (IPM) program was established to steer farmers away from routine pesticide use and to protect natural predators. The program was so successful that in 1986 the Indonesian government declared IPM the national pest-control strategy for rice. It banned 57 pesticides, ended many pesticide subsidies, and established an IPM education program for rice farmers. Pesticide use was cut by 65 percent, and rice yields increased 40 percent.

IPM successes are found closer to home as well. Because of the subtropical climate and the long, hot growing season, chemical use is more intensive in farming areas of south Florida than in most other farming areas in the U.S. In a major study of alternative agriculture in 1989, the National Academy of Sciences found that many growers in south Florida try to keep their pesticide use to a minimum by using IPM "scouting" services, in which professional pest scouts monitor crops carefully for pest populations and disease. The frequent monitoring of crops at every growing stage allows growers to catch pest or disease problems early on, making it possible to get more effective pest control with a much lower level of pesticides than would otherwise be needed. What's more, the need for routine follow-up applications is often eliminated. IPM programs are used for growing as much as

80 to 90 percent of the sweet corn in south Florida, and help cut pesticide use by 50 percent for insects and 25 percent for diseases. Similar programs have reduced insecticide use on Florida tomato farms by 21 percent. IPM has saved some farms as much as $400 per acre on pesticide costs.

(One cautionary note about any program to reduce pesticide use: Reducing pesticide *applications* isn't enough if that

Pesticide-reduction programs abroad

Since the 1980s, concerns about the environment, food and drinking-water safety, and the health of farmers and farm workers have led a few European countries to take serious steps to reduce pesticide use:

➤ The Netherlands has adopted a national program to require a 35-percent reduction in pesticide use by 1995, and at least a 50-percent reduction by the year 2000, along with reductions in pesticide pollution of the environment. Pesticide use in the Netherlands is currently among the most intensive in the world. Among other elements, the Dutch pesticide-reduction plan includes a massive commitment to research, education, and training, including training for anyone who buys, sells, or applies pesticides. Research is already under way on "model farms," and one such farm maintained crop yields even when it reduced pesticide use by 90 percent.

➤ In 1985, the Swedish government responded to broad public concern about the health impacts of pesticides by adopting a major pesticide-reduction program to cut agricultural pesticide use there by 50 percent by 1990. That goal has already been met, and the goal now is another 50-percent reduction by 1997. The program is a mixture of voluntary and regulatory measures, including compulsory training of spray operators, a wide variety of research programs, and technical assistance in preventive pest control. The plan also includes a subsidy program to help farmers making the transition to organic farming. One key reason for the success of the Swedish pesticide-reduction program so far has been the active cooperation of farmers and farmer organizations. According to the World Wildlife Fund, "From a position of initial suspicion, a majority of farmers have come to realize the economic and environmental advantages of the programme." ❧

means that more hazardous chemicals are being used in smaller quantities.)

IPM at your supermarket. IPM isn't just for small farms. In fact, IPM methods are being used by growers for some of the giant, well-known processors in the food industry. To keep pesticide residues out of jars of Gerber® baby food, for instance, the Gerber Products Company employs what it calls a "total systems approach" that starts with using IPM techniques on the farm to reduce or eliminate pesticide use in the first place. Gerber encourages its growers to use specialized crop rotations that help reduce the need for pesticides on peas, beans, and carrots. The company funds cooperative programs with universities to develop disease-resistant crop varieties, such as peach varieties that have increased resistance to bacterial leaf spot.

Gerber also takes advantage of climatic conditions in various parts of the country to reduce the need for pesticides. The green bean crop in Michigan is planted for harvest in August to avoid the need for pesticides to control the European Corn Borer, for example, and Gerber uses the spring spinach crop from Arkansas because fewer pesticides are applied there early in the year. Growers and suppliers across the country have to adhere to strict standards set by the company on planting, growing, and harvesting, and Gerber monitors the program's effectiveness with product testing.

Anyone who thinks reducing pesticides is somehow backward or old-fashioned need only look at an IPM program that Campbell Soup Company has developed for growing tomatoes that are processed into soup and other products. In Ohio, for instance, the company is using a computerized disease-forecast system for monitoring conditions that affect the potential for various diseases to develop. The program involves dividing the tomato production into separate zones, each with a 10-mile radius. An "Omni datapod" is placed in each zone to record hourly air temperature and leaf-wetness values. For each zone, "disease severity values" (DSVs) are calculated daily, and cumulative DSVs are recorded on an 800-phone line for growers to call. Instead of routinely spraying with fungicides, growers spray the crops only when DSVs exceed certain thresholds.

In the 1990–1991 growing season, growers who participated in the program used an average of four fungicide applications per acre, compared with an average of 6.5 for non-IPM growers. In 1992, growers who remained in the IPM program used an average of six fungicide applications per acre, compared with 9.5 applications by non-IPM growers. The reductions in fungicide use translated into a total savings of $382,000 for just those two years.

The computerized disease-forecast system is only one component of Campbell's IPM program. In all, the company contracts with growers farming some 30,000 acres to produce tomatoes in Ohio, California, and Mexico. All of the Mexican acreage and 95 percent of the Ohio acreage operate under a program of IPM developed by the company, and most of the California growers practice some form of IPM. In

IPM ant power

The San Francisco-based Bio-Integral Resource Center has documented a number of fascinating uses of non-chemical controls in China. For more than 1,600 years, for instance, Chinese farmers have used yellow citrus ants to control about 20 kinds of leaf-eating pests in citrus, mango, coffee, and other groves. In the spring, farmers gather citrus ant nests from the trees where they winter, and move the nests into the fruit trees. Water moats at the tree bases keep the ants from leaving, and bamboo poles in the branches form bridges for them to move among the trees in search of insect prey. Because chemical insecticides aren't used, beneficial lady beetles, spiders, and other predators survive to control smaller pests that the citrus ants miss. Trees with ants have almost two-thirds less fruit damage than trees where chemical pesticides are used. ❧

Mexico, an extensive insect-sampling program resulted in increased income, less crop damage, and 59 percent less pesticide usage by growers using IPM compared to non-IPM fields.

One problem for consumers is that it's impossible to tell whether the food at our local grocery stores was grown with many pesticides, or under a reduced-pesticide IPM program. Pesticide residues are invisible, and unlike with organically grown food, food grown using IPM methods isn't labeled as such. In addition, right now there is no official certification program for IPM farming. The cooperative extension service at the University of Massachusetts is trying to do something about that. Its IPM Education and Certification Project, a pilot program that was started in 1989, has created a statewide education program for growers and other farm professionals on the principles of IPM, and has developed a state IPM certification program with guidelines for several crops—including apples, strawberries, potatoes, sweet corn, cole crops, and cranberries. The next step will be a marketing program, to include consumer education and, eventually, an IPM label at the retail label. Starting in 1993, Massachusetts consumers may have some important new choices in their grocery stores, and IPM growers will get some well-deserved recognition.

Organically grown goodness

Larry Conklin grows tomatoes, squash, peppers, and sugar snap beans on his 14-acre farm in Newnan, Georgia. Across the country, Steve and Tom Pavich farm almost 2,500 acres in California and Arizona; in one year they produce well over a million boxes of grapes. North Dakota farmer Fred Kirschenmann raises grains and beans on about 2,000 acres. Liz Henderson and David Stern grow mostly asparagus, garlic, greens, root crops, and blueberries on their 60-acre farm in Rose, New York.

These growers market to small local stores, to grain companies, to major supermarkets and natural food store chains, and directly to people in their communities. What these farmers have in common is that they are farming *organically*—without synthetic chemical pesticides or fertilizers. *And they're successful.*

Not long ago, some "experts" might have scoffed at these farmers as impractical. When Larry Conklin told the agricultural extension agent that he was farming organically, the agent wished him luck. In 1980, organic farmers were encouraged when the USDA published a landmark report that acknowledged an increasing concern about the "extensive and sometimes excessive use of agricultural chemicals." The report surveyed organic farming methods, and observed that many of the soil and crop management practices followed by organic farmers were the "best management practices for controlling soil erosion, minimizing water pollution, and conserving energy." It concluded that large numbers of small farms in the U.S. could be converted to organic farming with little impact on the U.S. economy, and called for research and education programs to help address the needs and problems of organic farmers. Instead, shortly after the report was released, the Reagan Administration shut down the USDA office that produced it.

Organic farming took off anyway. The market share of food produced organically increased throughout the 1980s. In 1980, organic food sales were $174 million; in 1991, organic food sales totaled *$1.25 billion.* According to the

Organic Foods Production Association of North America (OFPANA), there are about 3,000 certified organic growers today—38 percent more than in 1990. Market experts predict that future growth in the organic food industry could be even more impressive—between 25 and 50 percent a year—and that organic food could represent 10 percent of the total food market in the U.S. by the year 2000. One thing that will no doubt help boost the organic industry will be a new, national standard and label for organically grown food that will be implemented in 1994.

The all-new organic label. Until now it has been hard for consumers to know just what it means when food is labeled "organic." Instead of one nationally recognized definition, half of the states have had their own laws or regulations governing organic food, and the standards have varied in stringency. Even "certified organic" has meant slightly different things, depending on the standards and verification procedures of the certifying agent, which can be a state agency or a private certifying organization.

All of that is about to change, however. In 1990, Congress passed legislation requiring a national standard and mandatory certification program for organically grown food. (The national organic certification program is the first of its kind—no other sustainable farming method has a national third-party certification requirement.) The certification program will be administered by the USDA, which will accredit state and private third-party certifying organizations. A new National Organic Standards Board, which is made up of organic farmers and food handlers as well as consumer and environmental representatives, is currently working to develop the specific standards for organic farming and processing.

The standards will require that organic food be grown primarily using natural materials (with very few exceptions for certain synthetic chemicals, which will be recommended by the board and finalized by the USDA). In addition, the board will recommend restricting the use of certain natural pesticides that have been used by some organic farmers even though they are very toxic. The new standards will cover meat, milk, and eggs as well as other foods, such as produce, grains, and processed food. Expanding the organic label to include all these foods is a breakthrough; up until now, for instance, the USDA hasn't allowed organically grown meat to be labeled and marketed as such, even when it has been certified organic.

Under the new law, organic farmers, food handlers, and processors will have to keep detailed records to show their compliance with the new standards. The records will form an audit trail through which a product can be traced from the consumer back to the farm and field. In addition, farms will be inspected annually and will be subject to additional unannounced inspections, soil testing, and testing for pesticide residues. "Organic" is a *production* claim, and not a residue-free content claim, but the new standards require that pesticide-residue levels in organic food be many times lower than may be present in conventionally grown food.

Food produced according to the new standards will carry USDA-approved "organically grown" labels that will identify the certifying agents. For consumers, the new law means that you will be able to shop for organic food with more confidence, knowing that anything labeled organic was grown according to uniform, strict production standards. What's more, you will be able to verify the authenticity of the organic claim yourself, simply by asking to see the documentation for any organic food sold in your local store.

The new organic certification law goes into effect in the fall of 1993. But the program was underfunded, so the painstaking process of defining the standards and the certifying process was slowed down, and parts of the law will not be implemented until 1994. Until then, when you buy organic food, keep looking for a "certified organic" label and the name of the third-party certifier—it's still your best insurance that the food was grown without synthetic chemicals.

The Pavich family

Organic farming: Some farmers' stories

➤ **Stephen, Tom, and Tonya Pavich, Pavich Family Farms.** If your supermarket stocks organic grapes, chances are they're grown by Pavich Family Farms. The world's largest organic table grape grower and shipper, Pavich is living proof that large-scale organic growing makes good *economic* sense as well as good environmental sense. The farm markets its grapes and other produce to natural food stores, the top 20 U.S. grocery chains, and outlets in London, Hong Kong, and Japan.

A family-run farm for more than 35 years, Pavich converted to organic methods in the early 1970s. That was when Stephen Pavich went back to his father's vineyards after getting his degree in viticulture (the study of grape-growing). As Pavich describes it, one day shortly after returning to the farm he was overcome by pesticide fumes when he was washing out a vineyard tank. He knew then that there was something wrong with farming with pesticides, and suddenly he looked at farming differently. "I questioned the safety of the workers, the pollution of the groundwater, even the overall quality of the crops," he says. So he started studying organic farming methods, and began the gradual process of eliminating synthetic pesticides and fertilizers. Conventional wisdom had it that it was impossible to eliminate chemicals in large-scale vineyards and still produce an attractive grape that consumers would buy. And it wasn't easy: Pavich's father once said their switch to organic farming was as difficult as putting together a 10,000-piece jigsaw puzzle where the top half is nothing but clear blue sky.

Today Pavich, his brother Tom, and Tom's wife Tonya manage Pavich Family Farms in California and Arizona, producing eight varieties of table grapes, four kinds of melons, and several varieties of vegetables, as well as organically grown

Networking down on the farm

For many farmers who have switched from chemical-intensive to more sustainable farming practices, one of the hardest parts has been having to go it alone. It wasn't just that these farmers couldn't find useful information about sustainable farming. It was also the isolation they felt as their attempts to reduce or cut out chemical use were met with everything from skepticism to hostility from many other farmers in their communities. That may be slowly changing, thanks to the increased interest on the part of many farmers in reducing chemical use—and thanks to some new farmers' networks that have been developed in almost every state to help out with the transition.

One of these is the Northeast Wisconsin Sustainable Farmers Network, a loose group of about 30 Wisconsin farmers who try out innovative methods on their farms and then share the results with one another—with the goal of improving environmental practices *and* the farms' profitability. The project has been a success, with some of the farmers reducing their fertilizer and herbicide use significantly and still maintaining yields at least as high as with conventional farming methods. The network has also learned, as coordinator John Bobbe has put it, that "blanket formulas and prescriptions will not work in sustainable agriculture. Farmers want specific information that can be provided for their individual farms and the way they choose to run their farming operations." ❧

cotton. They control pests mainly through soil-building techniques that strengthen the plants' immune systems, and by creating an environment that's conducive to beneficial predatory insects and parasites that attack unwanted pests. Instead of using herbicides, they control weeds mechanically through tillage and cover crops. They have replaced synthetic fertilizers with organic fertilizer and natural minerals. Their yields equal those of chemical-intensive farm operations, and Pavich's grapes look and taste as good as conventionally grown grapes—all without the huge costs (financial *and* environmental) of extensive pesticide use. "When I leave, I want this spot on the planet to be clean," says Stephen Pavich. "I don't want my child or any other child to worry about the side effects of the way we work the soil."

➤ **Norman Freestone, Ecology Sound Farms.** Norman Freestone raised citrus and other fruits conventionally for 12 years on his farm in the San Joaquin Valley, California, near the Sierra Foothills. Then in 1979 Freestone became seriously ill—among other things, he suffered from severe fatigue and anxiety, and lost 45 pounds. Doctors he went to couldn't diagnose the problem; some of them suggested that he get psychiatric treatment. Finally he saw a doctor in San Francisco who identified chemicals as the cause of his illness. The doctor advised him to quit farming. Instead, Freestone stopped using farm chemicals. "Switching to organic was a real effort for almost 10 years," he says. "I had no research to turn to from the universities. I just taught myself the system by trial and error—learning from my mistakes. Only in the last few years could I really consider myself a master of my own trade."

Today, Freestone successfully grows navel and valencia oranges, pomelos, plums, asian pears, kiwis, and persimmons on his 130-acre Ecology Sound Farm. He sells to natural food store chains in the U.S., exports to Japan (where there is a significant market for organic foods), and runs a small mail-order service.

Farmers: "Getting from A to Z"

Iowa farmer Sharon Thompson: "The vast majority of financially stressed farmers perceive that they cannot jump from one extreme to another. Conventional farmers perceive that the sustainable agriculture movement is asking them to make a complete change from step A to step Z all at one time. This has given the sustainable agriculture movement a bad name, and has built many walls between people. The perception needs to be changed to a practical, sensible approach of moving from step A to step B and then step C and then continuing to move toward Z." ❧

What organic farmers need, Freestone says, is real help from the government—in the form of economic incentives, such as tax credits or other incentives for converting to non-polluting forms of farming. Freestone has on his own attempted to further the cause of organic farming by challenging a system of federally approved marketing orders for navel oranges.

Marketing orders not only specify standards for size and appearance, but also volume restrictions—for instance, "prorates" limit how many oranges can go on the fresh market in any week. These limits are meant to protect the growers by preventing large price fluctuations and lengthening the marketing season, but in reality they benefit the large, conventional growers and end up hitting the organic growers hardest. (The prorates are determined by a committee of government officials, farmers, and packers—and large marketers such as Sunkist have the greatest influence in the committee.)

Normally, oranges in a grove ripen at about the same time, meaning that a large number of oranges are ready for the fresh market all at once. Conventional growers can use chemical growth regulators to keep oranges from dropping off the trees once they're ripe, but since organic growers don't use chemicals, their crops have to be harvested and marketed at once. If prorates prevent them from being sold to the fresh market, large numbers of oranges end up having to be sold for processing, which brings a lower price. In 1989, Freestone began petitioning the Agriculture Marketing Service, the branch of the USDA that oversees marketing orders, for an exemption from the volume restriction for organic oranges; his petition hasn't yet been ruled on.

For Freestone, farming organically simply makes sense from a health standpoint. "Pesticide poisoning is difficult to diagnose, and it is hard to get the medical community to take a stand," he says. "But our bodies are more sensitive than any laboratory. We need to use common sense. As far as I'm concerned, organic food is more than just food. It's a long-term, practical, and inexpensive form of preventive medicine."

➤ **Mel Coleman, Coleman Natural Meats.** It may seem incongruous for a cattle rancher to serve as an advisor to the Humane Society and repeatedly win awards from environmental organizations. But Mel Coleman is no ordinary cattleman. Coleman, a fourth-generation rancher in Saguache County, in southern Colorado, is a leading proponent of sustainable agriculture, reducing the use of farm chemicals, and humane treatment of livestock. That is saying a lot in today's meat industry, which inspired the term "factory farming."

Mel Coleman

Coleman practices and advocates cattle-raising methods that put the least stress on the environment and the livestock. Until the late 1970s, Coleman's family bred and raised calves—without using growth-stimulating hormones and antibiotics—and then sold the calves to cattle feeders who finished raising them using their own methods. In 1979, one of Coleman's daughters-in-law came up with the idea of raising the cattle themselves, and marketing the product as "natural" beef.

The marketing wasn't easy. Using the yellow pages as his guide, Coleman himself drove around to all the natural foods stores in Los Angeles to sell his natural beef. But the product found its niche, and today, Coleman Natural Meats are marketed widely throughout the country and in Japan. The company now works with more than 50 ranchers in six western states, who raise cattle and sheep according to Coleman standards—without antibiotics and growth stimulants, and using rotational grazing and other sustainable practices that protect and improve rangelands and prevent soil erosion. Ever the pioneer, in 1990 Coleman introduced the first nationally available, certified organic beef—from cattle fed only organically grown feed and raised on certified organic farmland.

Some of the marketing hitches have continued, though. First, Mel Coleman had to work hard to get the USDA to approve a "natural" label for his beef—but the USDA later watered down the definition so that it applies to any beef that has no artificial ingredients and is minimally processed, regardless of how the cattle were raised. And the USDA has not allowed meat to be labeled and marketed as "organic"—which will change when the new national organic standard and label go into effect.

Meanwhile, Coleman continues to break new ground. In early 1993, the company announced a new program of incentives for its ranchers to raise leaner beef by raising leaner animals—a notion that runs counter to the wasteful industry practice of raising fat cattle, and then trimming the fat after slaughter.

States taking the lead

Jim Hightower, the former Texas Commissioner of Agriculture, likes to quote a farmer friend of his as saying, "*Status quo* is Latin for 'this mess we're in.'" Under Hightower, the Texas Department of Agriculture made some impressive strides in shaking up the status quo and getting us *out* of the mess. In fact, several states have been innovators in supporting sustainable farming and consumers' right to know about pesticide use on food.

➤ Hightower's Department of Agriculture breathed new life into farmers' markets in the state, expanding the farmers' market program to include almost 2,500 new farmers, and increasing sales at the markets from $438,000 to $16 million. Hightower's department also established "Texas Certified Organic," the state's organic farming certification program, and helped organic farmers expand the markets for their crops.

➤ Several states have "linked deposit" programs, under which the states deposit money in private banks at lower interest rates than usual, on condition that the banks make low-interest loans available to farmers for alternative crop development—including farming with fewer or no chemicals.

➤ Iowa raises money for sustainable farming research and demonstration projects through a state tax on pesticides and nitrogen fertilizers.

➤ In 1986, Massachusetts Commissioner of Agriculture Gus Schumacher set up the country's first Farmers' Market Coupon Program, to provide food coupons to low-income families for buying fresh fruit and vegetables at farmers' markets and inner-city farm stands. Several other states followed suit, and the programs have been so successful that the federal government now provides matching funds. In 1992, at least 20 states used a combination of state, federal, and private funding to run these coupon programs. The programs help make nutritious, locally grown fresh produce available to low-income families, and help increase markets and earnings for local and regional farmers.

➤ The State of Maine passed two landmark food-labeling laws initiated by the Maine Organic Farmers and Gardeners Association (MOFGA)—one requiring retailers to label produce imported from countries that allow the use of pesticides banned in the U.S., and another requiring labeling of any produce that has been treated with post-harvest pesticides. ❧

Marketing a new food system

For farmers like these who have made the switch to sustainable agriculture, or the ones who are trying to make it, the key to success is finding steady markets for their products. Mel Coleman's first big break in marketing his natural beef came when he clinched a deal with Mrs. Gooch's, a natural foods store chain in Los Angeles. Norman Freestone's organic fruit is marketed by Mrs. Gooch's, as well as other local stores and chains such as the Maryland-based Fresh Fields. A whole new generation of supermarkets, distributors, and alternative markets (covered in the next chapter) is springing up to fill growers' need for a market, *and* consumers' need for better food choices. It is happening across the country:

➤ In Oklahoma, an innovative grocer is aggressively marketing organic food in his decidedly mainstream chain of grocery stores.

➤ In Atlanta, Georgia, a small local natural foods store named Sevanada helped revitalize an inner-city neighborhood.

➤ In San Francisco, a women-owned wholesale produce distributor, Veritable Vegetables, ships produce (grown by a very loyal following of farmers) across the country and around the world, and operates a retail produce dock that is frequented by the diverse local community.

While many of the stores offering choices like organic food are small to medium-sized independent natural foods stores, there's also a trend for sprawling, dazzling mega-markets that compete with the big conventional supermarkets in size and in the array of products they offer. In the early 1970s there were about a dozen of these natural foods supermarkets in the U.S. Now there are more than 150.

➤ **Fresh Fields: Maryland and Virginia.** Fresh Fields stores (there are six of them) are a far cry from the natural foods stores without freezer sections that Mel Coleman found when he first tried to market his natural beef. In fact, Fresh Fields was chosen as *Money Magazine*'s 1993 "store of the

Farmstands: Good food for the people, where the people live

"If organic produce is as good for you as most of us think, then it should be eaten by everybody—rich, poor, and everyone in between," says Mark Winne, executive director of the Hartford Food System, a non-profit organization in Hartford, Connecticut. "But if you're poor, reside in the inner-city, and want to purchase something clean and green, you face many barriers, especially a lack of education, money, and access to healthy food."

The Hartford Food System is doing something about those problems. Since 1990, the organization has run several community farmstands—"hybrids of rural roadside stands and urban farmers' markets," as Winne describes them. The farmstands make nutritious, locally grown produce available to low-income people right in their own communities. The stands operate once a week, from July through mid-October, and are located in convenient, well-trafficked areas in poor, inner-city neighborhoods in Hartford—making them easily accessible for elderly people and families with small children, who might have difficulty getting to farmers' markets.

Instead of having farmers come in and sell their produce directly (which might not be cost-effective for the farmers), the program buys produce wholesale from local farmers—many of them certified organic or practicing IPM—and resells it at reasonable rates at the farmstands. People shopping there can use food stamps, farmers' market coupons, or cash. And they can get more than just food—the farmstands also pass along free information about nutrition, food safety, and sustainable agriculture—including recipes for preparing unfamiliar vegetables, and bi-lingual brochures that describe problems with pesticides in food and the benefits of farming with fewer or no chemicals. At the end of the day, any leftover perishable food is distributed to local soup kitchens.

The Hartford program is creating a rare connection between the state's farmers and the inner-city poor. "We began with a vision: a local food economy jointly controlled by producers and consumers," says Winne. "We're building our own local food economy brick by brick." ❧

year." "Riding on the crest of eco-consciousness," the magazine says, "Fresh Fields serves up 'good for you' foods and products in snazzy supermarkets—for only 5% more than you would pay for the ordinary stuff." Fresh Fields offers hundreds of organically produced products, including goods from local and regional producers. Locally grown organic produce sold at Fresh Fields can be cheaper than non-organic produce at conventional supermarkets.

➤ **Whole Foods: Texas, Louisiana, California, North Carolina, Massachusetts, and Rhode Island.** Walk into any one of the 21 Whole Foods supermarkets (including Bread & Circus in Massachusetts and Rhode Island, and Wellspring Groceries in North Carolina), and you are immediately struck by the *choices* you have. In the produce section, for instance, you'll find not just organic food (certified—you can ask to see the documents), but an array of other possibilities. All of them are labeled and clearly explained—including "transitional" (grown in a way that meets organic growing standards, but hasn't completed the organic certification process), "pesticide-free" (grown without chemical pesticides, but chemical fertilizers may have been used), and even "IPM." Because of higher pesticide residues sometimes found on imported food, the store emphasizes domestically grown produce; for any imported produce, a country of origin flag is displayed.

A Whole Foods supermarket

In the meat department, there are "100% natural" tags that, unlike the USDA version, really mean that the animal was raised without antibiotics or synthetic growth hormones. And until the new national organic standard goes into effect, to get around the USDA restriction on labeling meat "organic," there are "organically fed" tags to indicate that meat was raised on feed grown without synthetic pesticides or fertilizers. Organic food is found throughout the store—including organic juices, organic sesame oils and olive oils, and cereals made from organically grown grains.

➤ **Pratt Foods, Oklahoma.** In contrast to the upscale feel of Fresh Fields and Whole Foods, Pratt Foods stores (there are eight of them in central Oklahoma—in rural, suburban, and city areas) look much more like everyday grocery stores, because they *are.* And Pratt Foods sells the everyday grocery store items, too, but with a big difference—owner and manager J.B. Pratt's zealous commitment to consumer education, and to providing a wider, sometimes surprising, range of choices. He tries various products at different stores: In one, non-toxic household cleaners share shelf space with the usual, chemical-laden kind, and cloth diaper supplies compete with disposables. He once offered nutritious boxed lunches for school kids in the same drive-in line that sold drivers coffee and doughnuts in the mornings.

Four of the stores have produce departments featuring organically grown food—over 50 items. In just three years, the portion of organic produce he sells has increased from 8 to 15 percent of his produce. "There's a way to be in the business of stocking organic produce, and then there's a *way,*" says Pratt. "I offer a wide variety—that's the secret of our success. You have to go into organic big, or not at all." When other conventional supermarkets gave up on the tiny organic sections in their produce departments, Pratt was determined to make his work, with a generous, attractive, and well-lit display at the front of the produce department; huge, splashy banners; full-page newspaper ads; television commercials; and lots of in-store information for consumers on the benefits of growing food organically. There were (and still are) brochures explaining how buying organic food is a vote for uncontaminated food and drinking water, conservation of soil and other resources, and the health of farm workers. There are flyers

J.B. Pratt

debunking myths about organic food (organic food is *not* hard to find, organic farming is *not* more expensive or less productive). "Those handouts disappear!" says Pratt. "People take them. I'm proud of that."

Pratt "grew up in the grocery business" and went on to attend medical school, but instead of practicing medicine ended up coming back to take over the family stores, to practice his own unique brand of supermarket management. With something like organic produce, he sees himself as an advocate. "I have to do more than just put choices out there for people to see. First, I have to sell my own people on the concept. All of my employees must understand what I'm doing and why—especially the managers and line people, since they talk to customers all the time. That's why the health food stores are so successful—their staffs are knowledgeable and customers go in there expecting to learn something."

Pratt admits that it's easier for him to initiate changes at his stores than it is for huge supermarket chains. "Smaller stores are closer to our supply sources," he says. "We can make changes quicker and be more flexible. We spend more time with our customers." Huge supermarket chains depend on volume and steady supply, and aren't geared to dealing with smaller suppliers and small quantities, he says. "A medium-sized guy like me can bob and weave more easily than the giants can," says Pratt. "It's tougher for them to change. But we can fix that."

That's right—we *can.* Now it's time for you to get out your grocery list, and start exerting your consumer power, because this is where *you* come in.

3

What You Can Do

Actions for Your Family, the Environment, and Our Future

We *can* change the way we grow our food. We can dramatically reduce our reliance on pesticides and other agricultural chemicals. Our farms can be productive and profitable—not just now, but for future generations. Farmers' and farm workers' health can be protected, and the quality of the soil and water supplies restored and preserved. Our food supply can be abundant and safe, and consumers can be given better food choices.

Right now, many farmers across the country are ready to cut back on the chemicals they use and instead turn to more sustainable farming methods—provided they get the support they need. You can help. This chapter and the next are full of ways for you to become involved in helping to change our food system—from simple steps to take at home or the grocery store, to organizing efforts that will take more commitment and time on your part, but with the potential for very satisfying results. Here's how:

➤ **Vote with your consumer dollars.** Start today to choose food on the basis of how it is grown. Every time you seek out organic food or food grown with reduced pesticides, for instance, you are casting a vote in support of safer food, and in favor of an approach to farming that respects the land, the environment, and future generations. Whenever you can, shop at stores or markets where you have clear choices: not just conventionally grown food, but certified organic or transitional organic, IPM-grown, certified "no detected residues," and locally grown food. If you don't find these choices available, ask for them. Talk to your neighbors and friends and encourage them to vote with their purses and wallets, too. We can create powerful consumer demand for food grown with fewer pesticides—to encourage many more farmers to adopt sustainable farming techniques, and many more retailers to make better food choices available.

➤ **Work for safer food choices.** You can help organize in your area for a farmers' market, farmstand, or community garden. At the same time, you might work with your supermarket to increase choices available to shoppers there—not just fresh produce, but processed foods and grains grown with fewer pesticides. You can set up a food co-op or other program to purchase food directly from farmers who are farming organically or reducing their pesticide use. You could work to get safer food in your child's day care center, and more.

➤ **Make your voice heard.** We have to let public officials know that we want our food grown safely and sustainably. As this book is being written, a major debate is heating up over how the federal government should regulate pesticides in food. We need a groundswell of public support for much more effective pesticide regulation, especially to protect children, *and* for a strong national commitment to sustainable agriculture.

You can make your voice heard in many ways: First, sign, stamp, and mail the preprinted postcards at the back of this book. These are targeted to the key congressional leaders who will be shaping new pesticide and farming legislation and guiding it through Congress. Second, copy and mail the letters we've included (or write your own) to President Clinton, EPA Administrator Carol Browner, and Agriculture Secretary Mike Espy. Third, urge others to write to these officials and to Congress in favor of sustainable farming policies. You can help get the word out by writing letters to the editor of your local newspaper, holding town meetings, getting your school board or community organizations to pass resolutions, and organizing letter-writing campaigns through your PTA, garden club, or other groups.

But meanwhile, you'll want to be feeding your family well...

Protecting your family today

➤ Start by elevating fruit and vegetables to their rightful place in your family's diet. They're sumptuous, packed with vitamins and fiber, and they can save lives: study after study has linked diets rich in plant foods with lower risks of heart disease and many kinds of cancer.

➤ Buy food grown sustainably whenever you can—it's your best protection against possible pesticide contamination. Many people are convinced that sustainably grown food tastes better than conventionally grown food. Farmers who are reducing or eliminating their pesticide use often select crop varieties for their taste and appropriateness to the local climate and growing season, while conventional produce varieties may be chosen mainly for their ability to *look* good in spite of harsh chemical treatments, rugged shipping conditions, and sometimes lengthy storage times. Help your family learn to appreciate the slight cosmetic imperfection in some sustainably grown produce—think of it as a badge of honor for having been grown without pesticides.

➤ Try to buy locally grown produce in season. It is almost always fresher, tastier, and closer to ripeness than other produce, and because it isn't being shipped long distances, it is less likely to have been treated with post-harvest pesticides—some of which are suspected carcinogens. Out-of-season produce is more likely to have been imported, possibly from a country with less stringent pesticide regulations than in the U.S.

➤ Wash all produce well. Use a vegetable scrub brush when appropriate. Adding just a few drops of a mild dishwashing soap to the water can help remove surface pesticides,

Eat your vegetables (and fruit)

Your mother was right . . . they are good for you. So good that the U.S. government has urged all Americans to eat at least five servings of fruit and vegetables every day—which is as powerful a reason as any to insist that they're grown safely!

To get the message out about the importance of fruit and vegetables, and to offer easy, practical ways to add them to the diet, the National Cancer Institute and the fruit and vegetable industry have developed a catchy "Five A Day" program. Groups representing more than 30,000 grocery stores have already joined in the campaign. To its credit, "Five A Day" has many corporations touting the virtues of vegetables to their employees, and has major food companies hawking, not junk food, but *fruit and vegetables* to kids. Hear, hear!

For a free copy of the brochure, "Eat More Fruits and Vegetables: Five A Day for Better Health," call 1–800–4–CANCER. ❧

but be sure to rinse thoroughly. Before washing, cut up things like broccoli and cauliflower, and when you buy conventionally grown produce, cut off the tops and leaves of celery, and remove the outer leaves of lettuce and cabbage, which may contain the most pesticide residues.

➤ Peel non-organic fruit and vegetables that are obviously waxed, to remove any surface pesticides that may be sealed in with the waxes. Just make sure you are getting plenty of fiber from other sources in your diet. (This is a good reason to buy organic food whenever possible—you'll get more dietary fiber, since you don't have to peel it to avoid fungicides that might be sealed in with waxes. Because organic produce isn't treated with post-harvest chemicals, you might find that the skins of some organic fruits and vegetables start to look dry or shriveled sooner than with conventional produce; this doesn't affect the taste or nutritional quality of the food.) Unfortunately, some pesticides are "systemic," and are contained inside the produce; they can't be removed by washing or peeling.

➤ Grow some of your own food if you can—without chemicals.

➤ Avoid using pesticides in your home, on your lawn, or around your garden. It's important to reduce your family's exposure to these chemicals as much as possible. (See resources, page 71.)

➤ If you're pregnant, take extra precautions to protect yourself from pesticide exposure. Avoid the kinds of fish that may contain high levels of pesticides—bluefish, striped bass, wild catfish, and fatty fish caught from polluted waters, such as salmon or lake trout from the Great Lakes. To avoid high levels of mercury, avoid swordfish, shark and fresh tuna.

Finding food that's grown sustainably

If you're already regularly buying food grown with fewer or no pesticides, then you probably have a favorite place to get it—a food co-op, say, or a farmers' market where you can shop weekly for fresh produce. But if you have never shopped for sustainably grown food, then you may not know the best places to look for it. The availability and sources vary according to where you live—here is a checklist to help you explore the possibilities:

✓ **Does your area have a farmers' market?** A farmers' market is one of your best bets for finding organic food or other food grown with fewer pesticides than usual. (And not only is it an excellent source of good food, it's good fun, too—the Union Square farmers' market in New York, with dozens of booths of exquisitely fresh and colorful bounty, is one of the most popular spots in the city on a Saturday afternoon in late summer and early fall.)

Farmers' markets provide consumers in cities and suburbs a rare chance to be in direct contact with the farmers who grow their food, and to talk to them about their farming methods. Farmers' markets support the farmers, since they eliminate the need for distributors—and distributors' profits. (And the prices are usually better than supermarket produce prices, for the same reason.) The food often tastes better, too, since it's fresher, and is likely to be picked closer to ripeness than food that gets shipped long distances. And it is less likely to have been treated with waxes and potentially hazardous post-harvest pesticides.

Fortunately, farmers' markets are becoming more common around the country—in cities, suburbs, and rural areas; the number of markets in the U.S. nearly doubled from 1980 to 1990. If you don't know whether there's a farmers' market in your area, try calling your county's agriculture extension agent, or your state agriculture department. If there *is* no farmers' market in your community, urge them to work with your town to set one up. You can put them in touch with farmers' groups in your state that may know of farmers interested in selling their produce at a farmers' market (see page 75).

✓ **Does your community have a food co-op?** These can range from buying clubs that operate out of someone's garage every week, to fully staffed storefront operations. They share many of the advantages of farmers' markets, including the availability of organic or IPM-grown food, the direct support for farmers, and the potential price savings for consumers. (For information on setting up a food co-op, see page 54.)

✓ **Is there a local natural foods store that stocks produce?** The best of these stores carry top-quality produce, especially locally or regionally grown produce, organic food, and IPM-grown food. Even the smallest natural foods stores often carry a good selection of organic cereals, grains, nuts, and

Coming soon to a store near you: Organic grains and processed foods

Organic food isn't just fresh produce. More and more organic products are hitting the market, including the *super*markets. Here is just a sampling of the organic bounty *(certified* organic) that you might soon find on your supermarket shelf (or freezer), from three superstars of the organic industry:

➤ **From Arrowhead Mills, Texas:** Organic adzuki beans, garbanzo beans, kidney beans, alfalfa seeds, bran flakes, corn flakes, yellow corn grits, seven-grain cereal, oat bran, barley flour, buckwheat flour, wild rice pancake and waffle mix, fresh pressed canola oil, fresh pressed safflower oil, creamy and crunchy peanut butter, and sesame tahini

➤ **From Earth's Best Baby Food, Vermont:** Organic baby foods: apples, apples and plums, pears and raspberries, corn and butternut squash, peas and brown rice, spinach and potatoes, carrots and parsnips, vegetable beef, sweet potato and chicken, macaroni and cheese, brown rice cereal, and blueberry yogurt breakfast

➤ **From Cascadian Farm, Washington:** Organic apple-grape juice, apricot conserves, wild blackberry conserves, raspberry-rhubarb conserves, all-fruit blackberry sorbet, frozen strawberries, frozen green beans, frozen sweet corn, frozen French fries, canned French-style green beans, baby sweet pickles and low-sodium sauerkraut . . . ❧

processed foods. Just be sure to shop as carefully at these stores as you do anywhere else, since not everything marketed at "health-food" stores is necessarily *healthy.* "Natural" high-fat food is no better for you than other high-fat food, for instance.

✓ **Is there a supermarket or grocery store in your area that stocks organic food or food grown with reduced pesticides?** Many of the supermarket chains that experimented for a while with organic produce gave up, but a few tenacious grocers, like J.B. Pratt in Oklahoma, persevered. Even if your store doesn't carry organic produce, there's no excuse for it *not*

"No detected residues"

You may have seen this claim on some foods. Through its "NutriClean" program, the California company Scientific Certification Systems certifies food as having "no detected residue," if it does not find any detectable pesticide residues above 0.05 parts per million (ppm) in testing of the food at the time of harvest. To be certified as NDR, growers have to provide records of all pesticides used; the company tests the food for those (and some other) pesticides. Currently, NutriClean-certified NDR food is being sold by four supermarket chains on the west coast—Raley's, Ralph's, Fred Meyers, and Andronico's.

NDR certification signals to the consumer that a grower has taken a significant step away from reliance on pesticides. NDR could be used as an enforcement tool at the end point of a regulatory system that required more sustainable farming. What we need is a clear national standard for what can be called "NDR," as well as strict record-keeping requirements (with public access to the records), plus effective testing for a broad range of pesticides.

Produce certified as NDR is probably better for consumers, farmers and farm workers, and the environment than conventionally grown produce, But NDR doesn't guarantee that a product is completely pesticide-free. Neither does an organic label. Both may contain trace amounts of pesticides, from irrigation water or rain, from the soil because of past pesticide use, or from sprays applied down the road. ❧

to stock organic grains, cereals, and canned and frozen foods. Don't hesitate to encourage your store's manager to stock these items—as well as fresh produce grown without pesticides or with fewer pesticides. Our tips on how to work with your supermarket follow below.

If you don't have access to any of these sources, there are many mail-order sources of organic food, ranging from farms that ship only kiwis, say, to companies with large full-color catalogs of products from organic juices and jams to meats. A couple of the biggest organic mail-order companies are Krystal Wharf Farms (717–549–8194), and Walnut Acres Organic Farms (800–433–3998). The book *Green Groceries,* by Jeanne Heifetz (HarperPerennial, 1992) is a useful, thorough guide to more than 275 organic food mail-order sources, big and small.

Increasing the choices in your supermarket

Supermarket managers care about what their customers want. Just look at your supermarket shelves—at the proliferation of low-fat products, for instance—for proof that the food industry and supermarkets can be responsive to consumers' concerns about healthy, safe food. If supermarket managers understand that enough people want food grown organically or with fewer pesticides, they will probably do their best to stock that food.

Working with your supermarket. While you can take on this project single-handedly, it will probably be easier (and more fun) if you do it with others. You can do this with a small, local group you organize (see "Organizing tip #1," page 79), or it can make a difference just to have some of your friends and neighbors working with you. And you may have more success with a medium-sized store that's independent or part of a smaller chain than with a store that's part of a huge chain, since the large chains may be less flexible in their buying practices.

The most important step is to get to know your supermarket manager and produce manager, and strike up an ongoing, cooperative dialogue with them. Don't view this as an adversarial relationship; after all, your goals are really the same—making high-quality, safe, and affordable food available for all the store's customers. There can be distinct advantages to knowing and winning over your produce manager—for instance, once your store stocks sustainably grown produce, your produce manager could let you know when weekly shipments will be coming in, so you can make a point of getting to the store and buying it while it's freshest. And if your produce manager is enthusiastic about stocking produce with no pesticides or reduced pesticides, he or she will find ways to do it creatively and successfully.

Before approaching the produce manager, figure out what you are trying to accomplish and what you're willing to do to achieve your goals. For instance, will you be satisfied if the store stocks just a few items of certified organically grown produce? A broader strategy might be to ask your store to agree to overall pesticide-reduction goals (see pages 88 and 89). You could also ask the store to stock grains and processed food such as baby food and frozen vegetables and fruits grown with no pesticides or fewer pesticides. And you should decide in advance whether you are willing to follow up with a letter-writing campaign to the corporate head of the supermarket chain if necessary to get action.

In your first meeting or discussion with the produce manager, try to learn about the produce that is already stocked at the store:

➤ Which fruits and vegetables are grown domestically, and where? Produce from very humid climates may be grown with heavier use of fungicides.

➤ Which produce is imported? Other countries often don't have the same level of pesticide controls as the U.S.—and in some cases we may even be importing produce that's been treated with chemicals that are banned here.

➤ Are there any plans to stock sustainably grown produce at the store? Has the store made successful or unsuccessful attempts at handling organic produce before? What were the problems?

By asking these questions, you're not just soliciting important information about the produce itself; you're also learning about the produce manager's attitude about pesticides. This can be a chance for you to clear up his or her misconceptions about organic or other sustainably grown produce, as well. Your produce manager may think that sustainably grown produce is inferior to conventional produce, for instance (it *isn't*), or may think that consumers won't buy fruit and vegetables with cosmetic blemishes (surveys have indicated that consumers are willing to buy less-than-perfect-looking produce when they are informed that it looks that way because fewer pesticides were used).

If the manager doesn't have answers to all your questions right away, give him or her a few days to get the information—just having a more informed produce manager is already a step in the right direction.

Next, you can talk with the store's manager, or write to the corporate management of the supermarket if it's part of a chain. Tell them you're interested in setting up a meeting to discuss the possibility of stocking food grown with fewer pesticides. At this meeting, you could ask the supermarket manager to sign the pesticide-reduction agreement. Offer to help write a letter to the store's suppliers asking them to comply with the agreement. You should be prepared to listen to any concerns and answer questions they have, and you may want to bring along a copy of this book to show what kinds of resources are available, including the organic wholesalers' directory on page 70 and the farmers' organizations listed on pages 74–78.

Understanding and overcoming the obstacles. Supermarkets may have legitimate problems that need to be worked out, but with time and the grocer's commitment, the obstacles can be overcome.

One problem that supermarket managers often cite is price. Organic food is often more expensive than conventionally grown food, for a number of reasons. For one thing, organic farming is more labor-intensive than conventional, chemical-intensive farming is. In addition, with organic food, you are paying upfront for proper stewardship of the land and the environment. With conventionally grown food, the costs of harm to the environment, health problems, and long-term wear and tear on farmland are not reflected in food prices—they are costs that hit society in other ways. What's more, most supermarkets sell only a small volume of organic food, and supermarkets rely on volume to keep prices down. But just as consumers are willing to put up with cosmetic imperfections in organic produce, surveys also show that consumers will pay more—within reason, of course—for produce grown without pesticides. And the experience of some grocers has shown that organic food can be sold at competitive prices. "If you stay in touch with your suppliers, you can make some excellent buys all through the year," says Oklahoma grocer J.B. Pratt.

There can be other, logistical problems for supermarkets as well—problems that conventional grocery stores aren't accustomed to and are labor-intensive to solve. There's the question of where to display organic produce, for instance—should it be mingled together with the conventional produce, so that organic carrots, say, are in a bin next to the conventional carrots? Or should all the organic produce be displayed in a conspicuous, specially marked area of its own? Either way, organic food needs to have its own labeling or packaging, not just for the consumer, but for the check-out clerk. If produce is bagged by the grower, in the field, it doesn't hold up as well in distribution. It's better to do that at the last minute, but that means higher labor costs on the part of the grocery store. Organic growers' and distributors' methods of cooling, packaging, and transporting produce have gotten much more sophisticated in the past few years, but since organic produce isn't sprayed with post-harvest pesticides to increase its shelf-life, unless there is good turn-around of organic produce, it won't stay fresh.

The produce manager is the key to solving most of these problems. A manager who is personally committed to giving customers the choice of organic food will "fire up the sales force," as Pratt puts it, and will promote it with attractive, well-maintained displays, advertising, and in-store signs and hand-outs. Signs or brochures could describe the environmental and safety benefits of organic farming, for instance, or point out that since ripeners and colorings aren't used on organic fruit, it might not look as bright and "picture-perfect" as conventionally grown fruit. Organic food samples could be offered to customers to show how good it *tastes.*

Other things supermarkets can do. In addition to signing the pesticide-reduction agreement and stocking and promoting sustainably grown produce, processed foods, and grains, there are other steps you can encourage your supermarket to take:

➤ **Buy from local and regional sources.** This just makes good sense from an energy and environmental standpoint. It also helps support local farmers, and means that the produce will be fresher and possibly less likely to be treated with post-harvest pesticides. If your store is part of a large chain, ask the corporate management to give individual stores more autonomy to buy from local and regional sources.

➤ **Label waxed produce.** Retailers are required by federal law to do this, but in the past few of them have. To increase compliance with the law, new FDA regulations will go into effect in May 1994, requiring shippers and retailers to prominently label any fresh fruits and vegetables that are treated with post-harvest wax or resin coatings. (The waxes themselves may be safe, but not all of them have been fully tested for safety. The main problem is that they can seal in post-harvest pesticides, or can be mixed with fungicides to help prevent mold. When these pesticides are used with waxes, they can't be washed off.)

➤ **Indicate which produce has been treated with post-harvest pesticides.** This is *not* required by the law or FDA regulations, but it would be particularly useful for consumers. It is easy for retailers to disclose this information, since shippers are required to declare it on shipping containers.

➤ **Indicate country of origin.** Again, retailers aren't required to do this, but it would be easy for them to, since that information is also included on shipping containers.

Alternative and lower-cost sources for safer food

Unless you insist on one-stop shopping or unless your produce manager is unusually committed, at this time you'll probably find better selections and prices on sustainably grown food at sources other than supermarkets. In addition to established community farmers' markets, there are a number of good alternatives, such as food co-ops, "Community-Supported Agriculture," and subscription purchasing programs, which eliminate distributors and can offer participants good prices and even hands-on farming experiences. Here's a description of these programs and some basic information on how to set them up. For resources with information on farmer organizations and wholesale sources of organic food, see pages 70 and 74–78.

Food co-ops. These allow consumers to bypass supermarkets and buy food in bulk directly from farmers and other suppliers. They can range from small groups of friends who purchase food together from wholesalers, to large stores that serve thousands of families. One food co-op in Brooklyn has 2,500 family-members, all of whom work a few hours a month at the co-op (there's child care for the kids). In return, members get good prices on good food, and have access to some of the best organic produce in town.

If there isn't an established store-front co-op in your area, you can start your own—a good way is with friends, fellow workers, or neighbors. Start by calling a meeting of anyone interested, to discuss the types of food you'd like to purchase, food suppliers, and the method of distribution (this can be anything from taking turns picking up purchases and delivering them to members, to having weekly distribution areas set up in someone's garage). Carefully select the foods that you would like to have available to purchase. Have members list the foods they use and would like the co-op to supply, and then compile a master list of these foods. A small group can then review this list to eliminate foods that are difficult to find or for which there wouldn't be enough demand. For information on sources for food, consult with other co-ops, local farmers and farmers' markets, and wholesale directories.

Whatever form your co-op takes, you'll need a co-ordinator, a cashier or treasurer, buyers, packers, and someone to make order forms and collate completed forms. These tasks are usually done on a rotating basis. Joining a co-op requires a commitment of time on your part—it isn't like making a weekly trip to the grocery store—but that commitment can pay off by putting you more in control of what food you can buy.

When co-ops fail, it's often because they lack an organized accounting system that works for their size. Basically, there are three different systems for handling money and purchases:

➤ *The "potluck" system,* in which all members contribute a fixed sum for every order. The

Profile of an activist: Deborah Schimberg, Southside Community Land Trust

Deborah Schimberg gave the gift of gardening to her community on the South Side of Providence, Rhode Island.

Schimberg had moved into the diverse, low-income neighborhood herself in 1980, after graduating from Brown University, across town. Schimberg was concerned about the children in the neighborhood playing in trash-filled abandoned lots—she decided to try to convert the lots into community gardens. And she did, with an initial donation of $5,000 from a local environmentalist. With the money, Schimberg bought several empty lots (some of them for $50 each), and formed the Southside Community Land Trust.

The land trust makes garden plots—each one 15 feet by 15 feet—available to local residents in exchange for $10 a year and the promise to help with garden cleanups twice a year. The gardeners grow their fruit and vegetables in elevated plots, to avoid possible contamination from any lead that might be in the soil. One woman grows crops like collards and cabbages that she remembers from her sharecropper days in Kentucky. A Laotian immigrant grows beans and coriander—the kind of crops he and his father grew in Laos.

Altogether, some 200 families have gardened next to one another in 11 community gardens across the South Side. Schimberg expanded into other community development projects as well, including a "City Farm" for inner-city children, and a greenhouse at a minimum-security prison for women, where prisoners grow organic salad greens that are sold to restaurants in the city. ❧

A young gardener

buyers take the accumulated money and purchase the food based on what the co-op as a whole has determined it needs. The food is then distributed in equal shares to all members.

➤ *The "preorder and prepaid" system,* in which each member family carefully fills out an order form and prepays for what it orders. Someone in the co-op collates the orders and tallies what the buyers need to purchase.

➤ *The "preorder and pay on delivery" system,* which is similar to the one above except that no money is collected at the time of the order. Instead, members each contribute a small deposit (say, $25) when they join the co-op. This money is then used to make the first purchase of food, and the deposit fund is replenished when members pay for their food upon delivery.

Community-Supported Agriculture. This is a good option if you live in or near a farming community. As the name implies, Community-Supported Agriculture (CSA) is a way for a community to help support local farming. It benefits the farmer, who is guaranteed a market for his or her harvest, and it gives you access to fresh produce every week during harvest periods—usually at a reasonable cost—and a chance to get to know the farmers and the land that produce your food. And you can choose a farmer who grows food organically or with minimal pesticide use.

CSAs were first introduced in the U.S. in the 1980s, and there are now about 100 CSA farms across the country. There are many different kinds of CSA programs, and CSAs can be tailored to your particular community's needs. But they all have a few common elements: willing consumers, a willing farmer, and a sense of community. At one farm, shareholders helped to replant an entire corn crop when insects destroyed the first crop; at another, the group built a greenhouse to allow the farm to have each year's first harvest earlier than usual.

In a CSA, consumers buy "shares" in a farm's harvest and usually participate in the farming process by helping out either during the harvest season or with other aspects of the farm operation. A season's cost for a share in a CSA can range from $200 to $1,000 per family, depending on the length of harvest season and the amount and variety of produce that's grown. There's a certain amount of risk involved; if the weather is bad for crops, for instance, the harvest will be smaller and so will your share in the harvest. Work at CSAs varies, but usually each member, or shareholder, participates in crop harvesting, bagging and distribution, or administrative paperwork. CSAs most often provide vegetables and fruit, but some groups have expanded to include items such as honey, meat, milk, and eggs. Some CSAs have started from scratch—that is, consumers and farmers have gotten together to lease or purchase land and start a farm. Obviously, though, an easier approach is to link up with an existing farm that's looking for a market for its products.

CSA farm

To set up a CSA, you'll need to start by doing some research. Make sure that you have enough interested and committed consumers. Survey your neighbors, parents' groups, church members, and other community residents to determine what kind of interest there is. Schedule a town meeting to discuss the idea and answer any questions that come up; it helps to include the farmer or farmers who

Profile of an activist: Roberta Willis, Connecticut Mothers & Others

It started one evening in 1989. Roberta Willis and a small group of mothers in a rural community in Connecticut were gathered to discuss what they fed their kids. At the meeting was their neighbor, Meryl Streep, who had just read an advance copy of a report by the Natural Resources Defense Council on the problems of pesticides in children's food. "Meryl wanted her friends and neighbors to know what the report said," says Willis. "She was planning to help draw national attention to the problems of pesticides, but she also wanted to see changes in her own town." Led by Willis, the women went on to form the first Mothers & Others group.

The women agreed that their main purpose would be educational. They fanned out to PTA meetings, mothers' clubs, and day care centers (where they distributed locally, sustainably grown fruit and juices), and held a public forum on pesticides—which was attended by more than 500 people. They also worked to convince the four grocery stores in the area to experiment with stocking organically grown fruits and vegetables, and arranged to get locally grown produce throughout the summer, through shares in a Community-Supported Agriculture (CSA) program.

What happened then was unexpected: the success of the group's CSA had the unintended effect of reducing the demand for organic produce in the supermarkets. "During the summer we all had our own gardens and shares in the CSA program," Willis says. "And in the winter many of our members joined co-ops." So the members decided to make an effort to get at least some of their produce from the supermarkets. "It's important that choices be available in mainstream stores," says Willis. "The point of our work isn't just to have access to safe, sustainably grown food ourselves, but to make sure that everyone does."

are interested in participating. Contact other CSA organizations and organic farm associations for advice on getting started.

Identify a core group of members who, together with the farmer, will draw up a budget, determine the way the harvest will be distributed, assign and manage volunteer tasks, and do the bookkeeping. If you're starting from scratch and need land, talk to city or town leaders about the possibility of a long-term land lease from the community with an option to buy. One carefully planted acre of land can provide produce for as many as 50 families. Some CSAs have purchased land and placed the acreage in a land trust to ensure that the land remains agricultural.

Subscription purchasing. Ward Sinclair used to be a reporter for *The Washington Post;* now he is an organic farmer, but he still visits the newspaper's offices once a week all summer long, to deliver bags of produce to his former colleagues. Sinclair's Flickerville Mountain Farm "subscription service" was the first in the country, and the idea has caught on with other farmers.

In a subscription service, consumers buy a set amount of produce from a single farmer, usually on a weekly pre-paid basis. The program is similar to a CSA, except that you're a "subscriber," not a shareholder, and you don't share in the risks of farming the way consumers do in a CSA.

A subscription program can be set up directly between a farmer and consumers, or between a farmer and a food co-op. The programs can vary quite a bit, depending mostly on the farmer and the size of the farm. Some farms with subscription programs offer a flat-fee membership that entitles consumers to purchase produce at a discount; some offer consumers the option to pick their own produce for an even greater discount. If you're interested in a subscription purchasing arrangement, speak to farmers at your farmers' market. Farmers there may already have such an arrangement with consumers; others may be willing to set it up.

A compelling benefit of both CSAs and subscription services is the direct connection between consumer and farmer. As Sinclair puts it, his subscribers have become better shoppers: "They pay more attention to the weather reports, they follow farm news more closely, and they have come to understand how they fit into the food chain." And, of course, they get good, fresh, organically grown produce.

Working with your restaurant

The Flickerville Mountain Farm subscribers aren't the only people who get to taste the farm's harvest. Sinclair also sells organic produce directly to some Washington, D.C., area restaurants. Across the country in San Francisco, Alice Waters has made an art out of buying the freshest, locally grown organic produce for her restaurant, *Chez Panisse.*

If you'd like to approach your favorite restaurant about using organic food, Waters recommends using diplomacy. She shares these tips:

➤ "Approach the owner gently, and inquire about where the produce comes from," she suggests. Say that you're asking because you want to know if the restaurant supports the people who are taking care of the land. "That's the most important point," says Waters. Then add that you're willing to pay the price for organically grown produce.

➤ You could begin by asking the restaurant to feature an organic salad, because organic lettuce is so widely available. "You won't be making an impossible demand if you start simply," she says.

➤ You could also connect the restaurant with a local farm. "That's how we began at *Chez Panisse,*" says Waters. "People just started bringing us things from their own backyards. They'd walk in with ten bunches of radishes and ask me if they could trade them for dinner. Then baskets of oranges grew to bushels of potatoes, and all these wonderful things—such as the most exquisite sorrel. We just wanted more and

Alice Waters

more. So we branched out with our own gardens, but found that wasn't effective. Finally, we learned how to reach out to all the farm communities around us. And that's the story of *Chez Panisse!*"

Organic produce is a big investment for a restaurant owner, Waters says. "But it all depends on how you look at the cost. In my mind, there's just no question what the value is. Organic costs more because of the effort to grow it. We have to start thinking about food in a different way. We have to care for it in a different way and use it in a different way. The rewards are beauty and flavor and profound nourishment. No one should ignore those things and no one *would* ignore them if all of our carrots came in two boxes. One would be labeled, 'sprayed with pesticides,' and the other would say, 'organically grown.' I think the restaurateur would choose the ones marked organic every time."

Getting safer food into day care centers

When it comes to feeding our children, our food choices become even more important. Although getting your child's school cafeteria to use organic, IPM, or locally grown food could be a tough sell, you can at least start to raise awareness about how your children's food is grown through your PTA and other parents' groups. (For instance, you could work to have your school board pass a resolution for safer food for kids—see page 81.)

Getting sustainably grown food into day care centers might be easier than working with school cafeterias, since the buying procedures are less centralized and more flexible. You could start by setting up a meeting between parents and the day-care center director to raise your concerns. A positive approach may work best—instead of emphasizing a fear of pesticides, stress that young children should have the safest, healthiest, and freshest food possible. A day care center could buy juice, fruit, and vegetables weekly from a farmers' market, directly from local farms if you live in an agricultural area, or through a food co-op.

Teaching the children well. It isn't enough just to feed kids well; we also have to teach them to appreciate where their food comes from. At Columbia University's Teachers College in New York City, one innovative curriculum program is helping kids understand the links between the earth, food, and health—and through a teacher's manual, is showing teachers how to teach those connections. "EarthFriends" teaches schoolchildren the whole "story" of food—growing, transporting, changing and packaging and buying, cooking and eating, and taking care of what's left over. The program includes a field trip to the EarthFriends classroom, which is set up with separate stations for each part of the food story. Children get to participate in all the stages—learning about growing food by planting seeds, for instance—all under a huge banner with the EarthFriends motto: "Good for us, good for the farmers, good for the planet." The children come out of the program knowing more than most adults probably do about where their food comes from, how it's grown, and how farming choices affect the environment.

In addition to fitting farming and food into a school curriculum, there are many other things to do with children to give them an understanding about the food system and the earth:

➤ Take your own kids on a farm tour, or encourage your children's school to organize a farm field trip. Your county's extension service or state department of agriculture may have information on farms in your state that have "pick-your-own" programs.

➤ Visit the farmers' market together. Unlike at a grocery store, children get to see the people who produce their food.

➤ Look forward to seasonal changes with your children—and the different fruit and vegetable treats that each season brings.

➤ Encourage your children's school to start a school garden, and explain the value of gardening without using chemicals. Or, start a garden at home with your child. Just a small plot of land can teach vivid lessons about insects, the weather, and nurturing soil and food.

Perhaps most important, you can show your children that *you* care about where your food comes from and the way it was grown. Let your kids see that you make deliberate choices when you buy food, and that you go out of your way, if necessary, to find food that is grown sustainably. Tell them the reason: that you want to support farmers who are taking care of the land, for their children—and for *all* of our children.

4

Tools for Organizing and Learning More

Agenda for a new food system

Mothers & Others for a Livable Planet is calling for a transformation of the U.S. food-production system away from chemical-intensive agriculture and toward sustainable farming methods that are safe, sensible, and economically viable. Our **Agenda for a new food system** will protect the environment, ensure the safety of our food, conserve soil and energy resources, and protect farmers and farm workers.

We urge others to join us in supporting these specific reforms:

Federal farm policy: Assist and reward farmers who convert to sustainable farming practices.

Right now, most farmers cannot afford to undertake the changes to sustainable practices without government support. We must remove the economic and other obstacles that currently discourage farmers from reducing or eliminating their use of pesticides, and instead help them realize the full benefits of switching to sustainable farming practices.

➤ Fully fund and implement existing programs that support sustainable agriculture and pesticide-use reduction—including the Organic Foods Production Program, the Integrated Crop Management Program, and the Sustainable Agriculture Research and Education Program.

➤ Revise commodity programs to eliminate penalties for farmers adopting crop rotation.

➤ Do not require farmers who participate in mandatory production controls to retire land, since this discourages crop rotation. Instead, allow farmers to decide how to meet their production goals over a two- to five-year period.

➤ Decouple income support from crop production to ensure that all farming systems and rotations are treated equitably.

➤ Implement targeted programs, such as crop insurance and farm credit, to reduce the financial risk farmers take when they switch to alternative farming systems.

➤ Establish national certification programs (including regulations and enforcement) for sustainable farming methods—Integrated Pest Management, no detected residues, and transitional organic.

➤ Tax synthetic pesticides at the wholesale level to fund incentives for farmers switching to sustainable farming practices.

➤ Demand that the USDA develop domestic and international markets for sustainably grown crops.

Research and extension: Give farmers the information, management assistance, and technologies they need to farm sustainably.

Existing extension programs in alternative farming are underfunded, and inadequate funding has postponed work in several important research areas, such as pest-monitoring practices, biological pest control, cover crops, crop rotation, and plant health and nutrition. We must redirect federal research and extension dollars toward sustainable farming research and development.

➤ Fully fund the long-term regional and multidisciplinary research, demonstration, and extension in the USDA's Sustainable Agriculture Research and Education Program. Focus the program on alternative farming practices and systems tailored for the major crop operations in each region of the country.

➤ Allocate substantial annual funding for sustainable farming research. Distribute the money through the USDA's competitive grants program to universities, private research institutions, foundations, industry, and farmers working with in cooperation with scientists.

Pesticide regulation: Ensure safer working conditions for farmers and farm workers, improve environmental quality, and reduce chemical residues in food and water.

The current federal pesticide regulatory program is not meeting its fundamental mandate to protect public health and the environment. After 40 years of federal regulation, farmers and farm workers are still at risk from pesticide use, the environment is still being degraded, and the public is exposed to pesticides in food at levels that may be unsafe—particularly for young children. We must overhaul the Federal Insecticide, Fungicide and Rodenticide Act and the federal Food, Drug and Cosmetic Act to protect public health and the environment.

➤ Make pesticide regulation consistent with other food-safety regulations by basing it on health risks only, rather than balancing the risks of a pesticide with its economic benefits. Establish a "negligible risk" standard, and prohibit the sale of foods with pesticide chemical residues unless the risk from those residues is negligible, meaning that it will not cause or contribute to adverse human health effects (including cancer, birth defects, and neurotoxicity) greater than one in a million over a lifetime.

➤ Set limits for pesticides in food that fully protect children and infants by taking into account their unique physiologies, consumption habits, and higher exposure to pesticides relative to body weight.

➤ Regulate old and new pesticides consistently. Base registration decisions on a single preventive and comprehensive regulatory approach that includes a cradle-to-grave analysis of the social, economic, and environmental costs.

➤ Phase out older, more hazardous pesticides, according to a schedule of adequate investment in the development and testing of safer alternative pest-control measures.

➤ Simplify procedures for taking a pesticide off the market if testing data is not adequate to support its safety, or where data demonstrates that the pesticide does not meet the negligible risk standard. Remove the requirement that the EPA buy back unused stocks of banned pesticides.

➤ In setting tolerances, account for inert ingredients and cumulative health effects of multiple pesticide exposures.

➤ Develop and adopt new inexpensive, easy-to-use residue-detection technologies as quickly as possible to permit more and better testing of food shipments for pesticide residues.

➤ Stop the export of pesticides banned or never legally registered for use in the U.S., and impose severe penalties on food importers whose shipments contain residues of those pesticides.

Food marketing: Change food purchasing and distribution practices to favor food grown sustainably; give consumers information about how their food is grown and treated.

Our food marketing system could play a crucial role in encouraging the widespread adoption of sustainable farming practices. But right now it works against sustainable farming, in part because of the centralized purchasing practices of large supermarket chains (which account for the majority of produce sales), and because stringent cosmetic quality standards put farmers who reduce or eliminate synthetic pesticides at a competitive disadvantage. The following are steps that the government and private industry should take to foster sustainable agriculture:

➤ Revise cosmetic and grading standards to emphasize the safety of food and de-emphasize cosmetic appearance. Involve consumers in designing the new standards, and take into account consumers' concerns, such as flavor, nutritional quality, and pesticide use.

➤ Develop information campaigns to explain to consumers the relationship of appearance to food quality and safety.

➤ Provide consumers with point-of-purchase information about where and how food has been grown, cosmetically altered, manufactured, handled, or genetically engineered so that they can make educated choices in the marketplace.

➤ Give individual supermarkets flexibility to buy locally and sustainably grown produce.

Resources

Books about sources for sustainably grown food:

Green Groceries: A Mail Order Guide to Organic Foods
by Jeanne Heifetz, HarperPerennial, 1992, $16.00
(Available in bookstores.)

The 1992 National Directory of Organic Wholesalers: Organic Food & Farm Supplies
California Action Network, 1992, $34.95
(Available from CAN, P.O. Box 464, Davis, CA 95617.)

The Humane Consumer and Producer Guide: Buying and Producing Farm Products for a Humane Sustainable Agriculture
International Alliance for Sustainable Agriculture (IASA) and The Humane Society of the United States, 1993, $15.00
(Available from IASA, 1701 University Avenue, S.E., Minneapolis, MN 55414.)

Books about agriculture policy, sustainable agriculture, and food safety:

Alternative Agriculture
National Research Council, National Academy Press, 1989, $24.95
(Available from National Academy Press, 2101 Constitution Avenue, N.W., Washington, DC 20055, 202–334–3313.)

Basic Formula to Create Community Supported Agriculture
Robyn Van En, CSA Indian Line Farm, 1990, $10.00
(Available from CSA Indian Line Farm, RR3 Box 85, Great Barrington, MA 02130. A listing of CSA farms in the U.S. is available for $5.00.)

Circle of Poison
by David Weir and Mark Schapiro, Institute for Food and Development Policy, 1981, $5.95
(Available in bookstores.)

Safe Food: Eating Wisely in a Risky World
by Michael F. Jacobson, Lisa Y. Lefferts, and Anne Witte Garland for the Center for Science in the Public Interest. Berkley, 1993, $4.99
(Available in bookstores.)

Books and magazines about reducing pesticide use at home and in your garden:

Basic Guide to Pesticides: Their Characteristics and Hazards
Shirley A. Briggs and the Rachel Carson Council, Taylor & Francis Publishers, 1992, $39.50
(Available from Taylor & Francis Publishers, 800–821–8312.)

The Child's Organic Garden
Lee Fryer and Leigh Bradford, Acropolis Books, 1991, $9.95
(Available in bookstores.)

Common-Sense Pest Control: Least Toxic Solutions for Your Home, Garden, Pets and Community
William Oklowski, Sheila Daar, Helga Oklowski, Taunton Press, 1991, $39.95
(Available in bookstores.)

Healthy Home, Healthy Kids
by Joyce M. Schoemaker and Charity Y. Vitale, Island Press, 1991, $12.95
(Available in bookstores.)

New Organic Grower's Four-Season Harvest
by Eliot Coleman, Chelsea Green Publishing, 1992, $17.95
(Available in bookstores.)

Organic Gardening Magazine
Rodale Press, 9 issues/year; $25
(Available from Rodale Press, 33 E. Minor Street, Emmaus, PA 18098, 215–967–5171.)

Rodale's Chemical-Free Yard & Garden
Rodale Press, 1991, $26.95
(Available from Rodale Press, see above.)

Toxics A to Z: A Guide to Everyday Pollution Hazards
John Harte, Chery Holdren, Richard Schneider, Christine Shirley, University of California Press, 1991, $20.00
(Available in bookstores.)

Consumer, environmental, and governmental organizations working on agriculture, pesticides, and/or food safety issues:

Most of these groups publish newsletters; you can write to them for their publication lists and information about their programs.

Bio-Integral Resource Center (BIRC)
P.O. Box 7414
Berkeley, CA 94707
510–524–2567

Center for Science in the Public Interest (CSPI)
1875 Connecticut Avenue, N.W., Suite 300
Washington, DC 20009
202–332–9110

Greenpeace
1436 U Street, N.W.
Washington, DC 20009
202–462–1177

Mothers & Others for a Livable Planet
(see tear-out form at back of this book)
40 West 20th Street
New York, New York 10011
212–727–4474

National Coalition Against the Misuse of Pesticides
(NCAMP)
701 E Street, S.E., Suite 200
Washington, DC 20003
202–543–5450

National Coalition for Integrated Pest Management
(NCIPM)
8000 Centre Park Drive, Suite 340
Austin, TX 78754
512–834–8762

Natural Resources Defense Council (NRDC)
71 Stevenson Street, Suite 1825
San Francisco, CA 94105
415–777–0220

Northwest Coalition for Alternatives to Pesticides (NCAP)
P.O. Box 1393
Eugene, OR 97440
503–344–5044

Pesticide Action Network (PAN)
965 Mission Street, Suite 514
San Francisco, CA 94103
415–541–9140

Public Voice for Food and Health Policy
1001 Connecticut Avenue, N.W., Suite 522
Washington, DC 20036
202–659–5930

Rachel Carson Council, Inc.
8940 Jones Mill Road
Chevy Chase, MD 20815
301–652–1877

USDA Extension Service
14th Street and Independence Avenue, S.W.
Washington, DC 20250
202–720–4111

National sustainable agriculture organizations:

Farm Verified Organic (FVO)
Route 1, Box 40A
Medina, ND 58467
701–486–3578

Organic Crop Improvement Association (OCIA)
3185 Township Road 179
Bellefontaine, OH 43311
513–592–4983

Organic Farming Research Foundation (OFRF)
P.O. Box 440
Santa Cruz, CA 95061
408–426–6606

Organic Food Production Association of North America (OFPANA)
P.O. Box 1078, 20 Federal Street, #3
Greenfield, MA 01301
413–774–7511

Organic Growers & Buyers Association
1405 Silver Lake Road
New Brighton, MN 55112
612–636–7933

Rodale Institute
222 Main Street
Emmaus, PA 18098
215–967–8405

The Wallace Institute for Alternative Agriculture
9200 Edmonston Road, Suite 117
Greenbelt, MD 20770
301–441–8777

State-based sustainable agriculture organizations:

These organizations may be able to guide you to farms in your state that have farm field days, pick-your-own programs, and subscription buying or Community-Supported Agriculture programs. They are also potential sources of information for supermarkets interested in stocking local and sustainably grown produce.

California Certified Organic Farmers (CCOF)
303 Potrero Street, Suite 51
Santa Cruz, CA 95060
408–423–2263

Northeast Organic Farming Association of Connecticut (NOFA/CT)
Box 386
Northford, CT 06472
203–484–2445

Georgia Organic Growers Association (GOGA)
P.O. Box 567661
Atlanta, GA 30356-6029
404–621–4642

Iowa Organic Growers & Buyers Association
P.O. Box 2935
Iowa City, IA 52244
319–337–3452

Kansas Organic Producers, Inc. (KOP)
P.O. Box 82
Whiting, KS 66552
913–873–3431

Maine Organic Farmers & Gardeners Association (MOFGA)
P.O. Box 2176, 283 Water Street
Augusta, ME 04338
207–622–3118

Maryland Organic Food and Farming Association
6201 Harley Road
Middletown, MD 21769
301–371–4814

Northeast Organic Farming Association of Massachusetts (NOFA/MA)
41 Sheldon Road
Barre, MA 01005
508–355–2853

Minnesota Food Association
2395 University Avenue, Room 309
St. Paul, MN 55114
612–644–2038

Mississippi Organic Growers Association (MOGA)
277 Hurricane Creek
Lumberton, MS 39455
601–796–4406

Northeast Organic Farming Association of New Hampshire (NOFA/NH)
White Farm, 150 Clinton Street
Concord, NH 03301
603–679–5718

Northeast Organic Farming Association of New Jersey (NOFA/NJ)
31 Titusmill Road
Pennington, NJ 08534
609–737–6848

Northeast Organic Farming Association of New York (NOFA/NY)
P.O. Box 21
South Butler, NY 13154
315–365–2299

Northern Plains Sustainable Agriculture Association
Box 36
Maida, ND 58255
701–256–2424

Ohio Ecological Food & Farm Association
65 Plymouth Street
Plymouth, OH 44865
419–687–7665

Oregon Tilth
P.O. Box 218
Tualatin, OR 97062
503–692–4877

Northeast Organic Farming Association of Rhode Island (NOFA/RI)
P.O. Box 83
Peace Dale, RI 02883
401–295–1030

Tennessee Alternative Growers Association
215 Morningside Lane
Liberty, TN 37095
615–563–2353

Texas Department of Agriculture Organic Program
P.O. 12847
Austin, TX 78711
512–475–1641

Northeast Organic Farming Association of Vermont (NOFA/VT)
15 Barre Street
Montpelier, VT 05602
802–229–4940

Mountain State Organic Buyers & Growers Association
c/o Keith Dix
Morgantown, WV 26505
304–296–3978; 304–293–4801

Wisconsin Rural Development Center
1406 Business Highway, 18/151 East
Mount Horeb, WI 53572
608–437–5971

Organizing tip #1: Starting a local organization

An organized group of concerned parents can be very effective at calling attention to problems with how our food is being grown, and working at the local level for solutions. Whatever you can do alone, you can do even more effectively if others join you. There's plenty for an organized group to do—for instance, convincing supermarkets to stock sustainably grown food, working with local officials and local farmers to set up farmers' markets or farm stands, and organizing letter-writing campaigns or other efforts to urge Congress to enact important reforms. Here are some tips for getting started:

➤ Start, quite simply, by talking to others. Recruit other people with similar concerns by talking to friends, neighbors, parents of children in your child's school, church members, and members of local civic organizations. Develop a core group of people who share your concerns.

➤ Educate yourselves. Start a resource file of articles and books at one member's house. Invite experts to address your group.

➤ Identify other organizations to work with you. Think of existing groups that are likely to be interested in this issue, such as environmental and consumer organizations, natural foods stores and food co-ops, farmers' groups, civic clubs, and PTAs and other organizations concerned with children's issues. Contact these groups to see if you can attend some of their meetings to talk about food, pesticides, and agriculture and to invite them to join with you in doing something about it.

➤ Hold an organizing meeting. Have hand-out material about the issue available, and pass around a sign-up sheet for people's names, addresses, phone numbers, and affiliations. Break up into committees to accomplish certain tasks—research, community outreach, fundraising, and public relations. Discuss undertaking projects such as attending local organizations' meetings to talk about food and agriculture issues, starting letter-writing campaigns to Congress, writing letters to the editor of the local newspaper, getting the school board or other institutions to pass a safe-food resolution, compiling information on local sources of sustainably grown produce, and approaching area supermarkets about stocking sustainably grown food and signing a pesticide-reduction agreement (see page 89).

Organizing tip #2: Holding a town meeting

A town meeting can be a good way for your organization to call community attention to a problem and to get the community interested in working together on solutions. Give yourselves plenty of time for planning—several months, if possible. Form committees and delegate responsibility for various tasks, such as location, speakers, invitations, and publicity.

➤ **Contacts.** Contact other groups in your area that have interests and goals similar to your group's. Try to get these groups interested in your events and involved in specific tasks. See if they will share their mailing lists for invitations or other informational mailings.

➤ **Funding.** Investigate different sources of local financial support. Try to have many of the necessities donated, including location, refreshments, and mailings.

➤ **Location.** Many public facilities, such as libraries, schools, and churches, allow groups to use their meeting rooms at no cost provided the meeting is free and open to the public.

➤ **Invitations.** Invite all local officials in your area, including members of Congress. Also invite civic and environmental leaders, supermarket managers and produce managers, farmers, and farmer organizations.

➤ **Promotion.** Notify all the local media about your event, and invite them to attend. Try to get advance publicity in order to generate a good turnout at the meeting. Ask radio stations to broadcast a public service announcement (PSA) announcing the event. Post flyers at grocery stores, day care centers, libraries, and other public places.

➤ **Speakers.** Assemble a panel of experts from your local area. Brief them on the format of the meeting, the audience, and their time allotment for speaking. You could assemble a panel including organic or IPM farmers, representatives from parents' groups, consumer and environmental organizations, and pediatricians.

➤ **Format.** Choose a moderator who can be comfortable keeping speakers on track and within speaking time limits, and directing questions from the audience. Allow plenty of time for a question and answer period after the panel members have made their presentations—this is often the liveliest part of the forum.

Organizing tip #3: Passing a resolution

Organizations such as school boards often use resolutions as a way to take public stands on important issues. Convincing your own school board or other local group to pass a safe-food resolution can help raise your community's awareness about the issue and to get community leaders to address the question of how our food is grown.

Here's how you do it (you can follow this same process to get other groups—such as garden clubs, PTAs, or civic organizations—to pass safe-food resolutions as well):

➤ Begin by selecting one member who is likely to be receptive to your proposal; ask to meet with him or her, and when you meet, take along information on the issue of food and agriculture, such as a copy of this book. End your discussion by encouraging the board member to introduce a resolution to the full board. (There is a sample resolution on page 90.)

➤ Be sure to have lots of supporters attend the meeting at which the resolution will be discussed and voted on. Invite local newspapers and TV and radio stations as well, and appoint an articulate spokesperson who will be available to explain your position to the media. To support your effort, you may want to have signed petitions from parents calling for passage of the resolution.

➤ Once the resolution is passed, get word about it out to the public. Send copies of the resolution to all the local media, issue a press release explaining the board's action, and, if you'd like, include information on other activities that your group has planned. If you can afford to, you might consider taking out an ad in your local paper to reprint the resolution in full.

Organizing tip #4: Letter-writing

Elected officials pay attention to mail they receive from constituents. Although phone calls to Congress can be effective when a measure faces a close vote, in general more weight is given to written letters. The best way to write to government and elected officials is to express your personal concerns and be specific about what action you want taken. Include your name and address, and type or write your letter legibly. If you are asking for action on a specific bill, you can ask for a response to your letter.

On pages 85-88 we've included sample letters to President Clinton, the EPA, the FDA, and your local supermarket. In addition, there are tear-out postcards to mail in to key congressional members.

Here are other letters you could write:

➤ **To Vice President Al Gore** emphasizing the environmental impact of pesticide use.

➤ **To your own U.S. Senators and Representative,** asking them to work to establish stricter controls on pesticides in food, increased research on sustainable farming, and incentives to farmers who switch to sustainable farming methods.

➤ **To the editor of your local newspaper** (the letters page is one of the most popular parts of the newspaper, so it's a good way to call attention to an issue).

➤ **To state agricultural officials** to see what programs your state has to promote sustainable agriculture.

➤ **To state legislators** to encourage them to enact state-based sustainable agriculture programs, and to ask them to put pressure on Congress and federal officials for important reforms.

Sample letters, supermarket pesticide-reduction agreement, and safe-food resolution

The next several pages contain sample letters to President Clinton, the EPA Administrator, and the Secretary of Agriculture, as well as a sample letter to your supermarket, a supermarket pesticide-reduction agreement, and sample safe-food resolution. (The supermarket pesticide-reduction agreement is adapted from a supermarket agreement that was developed by the Consumer Pesticide Project of the National Toxics Campaign.) You can photocopy these pages to mail or to use in your local organizing, or use these as samples to inspire you to write your own.

President Bill Clinton
The White House
1600 Pennsylvania Ave., N.W.
Washington, D.C. 20500

Dear President Clinton:

My family wants the safest food possible. I care about the way that food is grown, and am concerned about the overuse of pesticides in farming today. America needs a safe and sustainable food production system—one that protects the environment and public health, and keeps all farmers farming. I ask that you make a strong national commitment to sustainable agriculture as the top agricultural priority of the country, and that you adopt strict pesticide regulations to fully protect our children from pesticides in their food.

We must change the way our food is grown. Right now, farm pesticides are contaminating the food supply, hurting the environment, and threatening the health of farmers and farm workers. Fortunately, tens of thousands of farmers across the country are already showing us that sustainable farming is both possible and practical. With ingenuity, innovation, and labor, they are reducing their use of synthetic pesticides and cutting their costs—while maintaining their farms' productivity and protecting the environment.

Sustainable agriculture offers the promise of a safer food supply and an environmentally sound growing system. Along with related businesses, it is a key to job growth in rural America, and to creating new environmental technologies. But because the transition to sustainable farming can be financially risky, we have to offer farmers the technical and financial support they need to make the switch. If this means that safer food costs a little more, I'm willing to pay for it.

An investment in sustainable farming is an investment in our environment and our children's future. Please lead the country to make that investment.

The Honorable Carol Browner
Administrator
U.S. Environmental Protection Agency
401 M Street, S.W.
Washington, D.C. 20460

Dear Ms. Browner:

I am very concerned about the safety of the food I feed my family, and about the way pesticides are being used and regulated in the U.S. today. The overuse of pesticides in farming is causing serious health problems for farmers and farm workers, damaging the environment, and compromising the safety of the food supply.

We must overhaul our regulation of pesticides to ensure that the environment and the health of farmers and farm workers is protected, and that our food is safe for everyone—especially our children. We need a pest-control policy that fosters the development of safer pesticides and non-chemical pest-control methods to replace hazardous pesticides in use now. And we need strict limits for pesticides in food that fully protect young children, who are probably more vulnerable than adults to pesticides' toxic effects. The regulation of pesticides in food must be based, not on cost/benefit, but on health (and must take into account *all* health risks, including cancer, neurotoxicity, and birth defects).

I urge you to lead the EPA to heed its mandate to protect public health and the environment—through a comprehensive, preventive approach to regulating pesticides.

The Honorable Mike Espy
Secretary of Agriculture
U.S. Department of Agriculture
14th and Independence Avenue, S.W.
Washington, D.C. 20250

Dear Secretary Espy:

I want to feed my family a safe and healthy diet, and I am concerned about the overuse of pesticides in growing our nation's food. I ask you to declare as your number-one priority the transition to a safe and sustainable food production system—one that relies less on synthetic chemicals and protects farmers and farm workers, the environment, and consumers. Already, tens of thousands of farmers around the country are demonstrating that this is possible by reducing or eliminating their use of pesticides. They are maintaining their crop yields, saving their farms and keeping them profitable, and conserving soil and energy resources. It's time to make it possible for many more farmers to join them in farming sustainably.

As Secretary of Agriculture, you are in a position to oversee the transformation of the USDA—with a full shift in priorities to encourage, assist, and reward farmers who switch to sustainable farming methods. The USDA needs to redirect its current research and extension-program funding from conventional, chemical-intensive agriculture to support sustainable agriculture. And your department should lead the way to a new food-marketing system that emphasizes safety and sustainability.

Let's put good farm stewardship first. Sustainable agriculture offers the promise of a safer food supply and an environmentally sound growing system. Along with related businesses, it is a key to job growth in rural America, and to creating new environmental technologies. An investment in sustainable farming is an investment in our environment, the future of farming in the U.S., and our children's future. Please lead the country to make that investment.

Supermarket letter

I want to feed my family a safe and healthy diet, and I am concerned about the overuse of pesticides in growing food in our country. I shop at your store, and urge you to take steps to ensure that the food that you sell is the safest possible. In particular, I call on you to agree to the pesticide-reduction goals set out in the enclosed supermarket pesticide-reduction agreement.

In addition, I ask that you give your customers clear choices in the food we buy, by labeling produce to indicate where it was grown and what pesticides (if any) were used in growing it or treating it after harvest. Federal law already requires all retailers to label produce that has been treated with waxes; I urge you to begin right away to comply with that law.

These steps will help your customers to make informed buying decisions. By respecting our need for good information and our concern about the safety of our food, you will be building our loyalty to your store. And together, we will be increasing the demand for food that is grown safely and sustainably.

Supermarket pesticide-reduction agreement

We, the undersigned supermarket representatives, agree to adopt these pesticide reduction goals in our store(s). We believe our customers deserve the best fresh fruits and vegetables, grown with the fewest and safest pesticides, to ensure a cleaner environment and the safest possible food:

1. Whenever possible, we will carry certified organic produce, grains, and processed food, and produce certified to contain no detected pesticide residues. We will also seek out food grown using Integrated Pest Management, and whenever possible will purchase produce from local farmers. We are committed to a phase-in of our entire produce section to certified organic or no detected residues, certified IPM-grown, and locally grown produce by January 1, 1998.

2. In order to identify growers who use the fewest and safest pesticides, our stores will request all suppliers of fresh fruits and vegetables to fully disclose all pesticides used to grow our produce. This will include disclosure of all insecticides, herbicides, fungicides, and other pest-control agents. Growers will be required to provide records to verify all claims of no detected residues, IPM, organic, or transitional organic. We will in turn make this information on pesticide use and verification claims available to our customers. We urge our suppliers not to sell produce grown with pesticides lacking a practical detection method.

3. We will actively discourage our suppliers from selling produce grown with any known or probable cancer-causing pesticides identified by the EPA. Our stores will not knowingly sell any fresh fruits or vegetables that have been treated with cancer-causing agents. In addition, we will not knowingly sell any produce treated with pesticides that are toxic to the nervous system, immune system, or reproductive system, unless they can meet a standard of no detected residue. This policy will be enacted no later than January 1, 1998.

4. We will support national, state, and local legislation and regulations to promote these pesticide reduction goals.

Safe-food resolution

Since children consume proportionally more fruits and vegetables than adults do, which increases their exposure to pesticide residues at a time in their lives when they may be most vulnerable to pesticides' hazards;

Since current "tolerances," or legal limits, for many pesticide residues in food were set for adults without considering children's consumption patterns or their greater vulnerability, and do not adequately protect children from the toxic effects of certain chemicals in certain crops;

Since the food supply will be best protected if growers reduce their use of pesticides; and

Since government studies indicate that pesticide use could be reduced significantly without adversely affecting yields or the quality of produce, but since farmers who are interested in reducing pesticide use currently face financial and other obstacles;

We therefore

1. Support legislation directing the federal government to set pesticide tolerances in food at levels that adequately protect children, taking into account children's eating patterns, higher rates of food consumption on a body-weight basis, and special sensitivity to certain toxins; and

2. Recommend that Congress actively support growers in switching from conventional, chemical-intensive agriculture to reduced-pesticide techniques.

NOTES

NOTES

The Honorable Patrick Leahy
Chairman, Senate Committee on Agriculture, Nutrition,
and Forestry
328A Russell Senate Office Building
Washington, D.C. 20510

The Honorable E. Kika de la Garza
Chairman, House Committee on Agriculture
1301 Longworth House Office Building
Washington, D.C. 20515

Dear Chairman Leahy:

My family needs safe food. I care about the way that food is grown, and am concerned about the overuse of pesticides in farming today.

We need to transform our food production system from chemical-intensive farming to sustainable farming that protects the environment and public health. We need a major shift in government policies to give farmers the technical and financial support they need to eliminate or reduce their use of pesticides. Many thousands of farmers are already showing us the way and proving that this sustainable farming is possible and profitable. Now Congress must make sustainable agriculture the top agricultural priority of the country.

Signed ____

Address ____

Dear Chairman de la Garza:

My family needs safe food. I care about the way that food is grown, and am concerned about the overuse of pesticides in farming today.

We need to transform our food production system from chemical-intensive farming to sustainable farming that protects the environment and public health. We need a major shift in government policies to give farmers the technical and financial support they need to eliminate or reduce their use of pesticides. Many thousands of farmers are already showing us the way and proving that this sustainable farming is possible and profitable. Now Congress must make sustainable agriculture the top agricultural priority of the country.

Signed ____

Address ____

The Honorable Henry Waxman
Chairman, House Subcommittee on Health and the
Environment
2408 Rayburn House Office Building
Washington, D.C. 20515

The Honorable Edward M. Kennedy
Senate Committee on Health and Human Resources
315 Russell Senate Office Building
Washington, D.C. 20510

Dear Chairman Waxman:

I am very concerned about the safety of the food that I feed my family, and about the regulation of pesticides that are used in growing that food. We need a comprehensive, preventive approach to pesticide regulation, and food safety standards that are based not on *cost/benefit,* but on *health.* These health-based standards should be designed to fully protect children, who are probably more vulnerable than adults to the toxic effects of pesticides.

We have to provide our farmers with the safest possible tools to grow our food. Congress must overhaul the way we regulate pesticides so that the more hazardous pesticides in use today are phased out and replaced with safer pest-control methods—for the sake of our children and our environment.

Signed ____

Address ____

Dear Senator Kennedy:

I am very concerned about the safety of the food that I feed my family, and about the regulation of pesticides that are used in growing that food. We need a comprehensive, preventive approach to pesticide regulation, and food safety standards that are based not on *cost/benefit,* but on *health.* These health-based standards should be designed to fully protect children, who are probably more vulnerable than adults to the toxic effects of pesticides.

We have to provide our farmers with the safest possible tools to grow our food. Congress must overhaul the way we regulate pesticides so that the more hazardous pesticides in use today are phased out and replaced with safer pest-control methods—for the sake of our children and our environment.

Signed ____

Address ____

7 steps
to protecting your family from pesticides in food

1. Feed your family plenty of fruit and vegetables—and whenever you can, buy food grown without pesticides or with fewer pesticides. Look for labels indicating that food is certified organic or transitional organic, grown using Integrated Pest Management, or certified to contain no detected residues.

2. Buy locally grown produce whenever possible. Because it isn't shipped long distances, it is less likely to have been treated with post-harvest pesticides. Farmers' markets or farmstands are good sources for locally grown food.

3. Try to avoid imported produce. Out-of-season produce is more likely to have been imported, possibly from a country with less stringent pesticide regulations than in the U.S.

4. Wash all produce well. Use a vegetable scrub brush when appropriate. Adding a few drops of a mild dishwashing soap to the water can help remove surface pesticides on conventionally grown produce, but be sure to rinse thoroughly.

5. Peel non-organic fruits and vegetables that are obviously waxed, to remove any surface pesticides that may be sealed in with the waxes. Be sure you're getting plenty of fiber from other sources in your diet.

6. Grow some of your own food if you can—without chemicals. Avoid using pesticides in your home or on your lawn. It's important to reduce your family's exposure to these chemicals as much as possible.

7. Speak out for a safer food production system. Ask your supermarket manager to stock sustainably grown food. Write to your congressional representatives to urge them to pass better pesticide controls and to make sustainable farming our nation's top agricultural priority.

Mothers & Others for a Livable Planet
40 West 20th Street, New York, NY 10011